Rather than resting on his well-earned laurels as one of cultural studies' leading theorists, Tony Bennett has produced a tour-de-force. *Making Culture, Changing Society* elegantly and rigorously charts new paths for social theory, and raises new challenges for cultural studies. It will be absolutely essential reading for scholars grappling with the new interest in the material and temporal structures of society.

—*Elspeth Probyn, Professor of Gender and Cultural Studies,*
University of Sydney

Making Culture, Changing Society is yet another very original, insightful and indispensable intervention by Tony Bennett. His newest book examines how certain Modern, post-Enlightenment truths about 'culture' were shaped and deployed during key moments in the history of institutions and disciplines such as anthropology and aesthetics. In this way, the book provides a compelling 'genealogy' of how a Modern reasoning and study about culture has operated as a form of social control, and it poses questions with which anyone who currently is engaged in the cultural sciences, cultural criticism, or cultural studies needs to grapple, because in many ways we continue to live within the history of institutions and definitions of culture which Bennett's book charts.

—*James Hay, Professor and Director of Graduate Studies, Institute of*
Communications Research at the University of Illinois

Announcements of the end of culture are premature! With customary erudition and extraordinary historical breadth, Tony Bennett takes us to the heart of the culture complex, navigating the labyrinth of practices, materials and sites – from studios and workshops to museums and exhibitions – that have made culture such a pervasive and flexible instrument of governing.

—*Georgina Born, Professor of Music and Anthropology,*
University of Oxford

Making Culture, Changing Society

Making Culture, Changing Society proposes a challenging new account of the relations between culture and society, focused on how particular forms of cultural knowledge and expertise work on, order and transform society. Examining these forms of culture's action on the social as aspects of a historically distinctive ensemble of cultural institutions, it considers the diverse ways in which culture has been produced and mobilised as a resource for governing populations.

These concerns are illustrated in detailed case studies of how anthropological conceptions of the relations between race and culture have shaped – and been shaped by – the relationships between museums, fieldwork and governmental programmes in early twentieth-century France and Australia. These are complemented by a closely argued account of the relations between aesthetics and governance that, in contrast to conventional approaches, interprets the historical emergence of the autonomy of the aesthetic as vastly expanding the range of art's social uses.

In pursuing these concerns, particular attention is given to the role that the cultural disciplines have played in making up and distributing the freedoms through which modern forms of liberal government operate. An examination of the place that has been accorded habit as a route into the regulation of conduct within liberal social, cultural and political thought brings these questions into sharp focus. The book will be of interest to students and scholars of sociology, cultural studies, media studies, anthropology, museum and heritage studies, history, art history and cultural policy studies.

Tony Bennett is Research Professor in Social and Cultural Theory in the Institute for Culture and Society at the University of Western Sydney. His work has had a defining influence on contemporary debates in cultural studies and cultural sociology. *Making Culture, Changing Society* builds on and extends his distinctive perspective on the relations between culture and society developed in his *The Birth of the Museum*; *Culture: A Reformer's Science*; and *Pasts Beyond Memory*.

Culture, Economy and the Social

A new series from CRESC – the ESRC Centre for Research on Socio-cultural Change

The *Culture, Economy and the Social* series is committed to innovative contemporary, comparative and historical work on the relations between social, cultural and economic change. It publishes empirically based research that is theoretically informed, that critically examines the ways in which social, cultural and economic change is framed and made visible, and that is attentive to perspectives that tend to be ignored or side-lined by grand theorising or epochal accounts of social change. The series addresses the diverse manifestations of contemporary capitalism, and considers the various ways in which the 'social', 'the cultural' and 'the economic' are apprehended as tangible sites of value and practice. It is explicitly comparative, publishing books that work across disciplinary perspectives, cross-culturally or across different historical periods.

The series is actively engaged in the analysis of the different theoretical traditions that have contributed to the development of the 'cultural turn' with a view to clarifying where these approaches converge and where they diverge on a particular issue. It is equally concerned to explore the new critical agendas emerging from current critiques of the cultural turn: those associated with the descriptive turn for example. Our commitment to interdisciplinarity thus aims at enriching theoretical and methodological discussion, building awareness of the common ground that has emerged in the past decade, and thinking through what is at stake in those approaches that resist integration to a common analytical model.

Series titles include:

The Media and Social Theory (2008)
Edited by David Hesmondhalgh and Jason Toynbee

Culture, Class, Distinction (2009)
Tony Bennett, Mike Savage, Elizabeth Bortolaia Silva, Alan Warde, Modesto Gayo-Cal and David Wright

Material Powers (2010)
Edited by Tony Bennett and Patrick Joyce

The Social after Gabriel Tarde: Debates and Assessments (2010)
Edited by Matei Candea

Cultural Analysis and Bourdieu's Legacy (2010)
Edited by Elizabeth Silva and Alan Ward

Milk, Modernity and the Making of the Human (2010)
Richie Nimmo

Creative Labour: Media Work in Three Cultural Industries (2010)
Edited by David Hesmondhalgh and Sarah Baker

Migrating Music (2011)
Edited by Jason Toynbee and Byron Dueck

Sport and the Transformation of Modern Europe: States, Media and Markets 1950–2010 (2011)
Edited by Alan Tomlinson, Christopher Young and Richard Holt

Inventive Methods: The Happening of the Social (2012)
Edited by Celia Lury and Nina Wakeford

Understanding Sport: A Socio-Cultural Analysis (2012)
By John Horne, Alan Tomlinson, Garry Whannel and Kath Woodward

Shanghai Expo: An International Forum on the Future of Cities (2012)
Edited by Tim Winter

Diasporas and Diplomacy: Cosmopolitan Contact Zones at the BBC World Service (1932–2012) (2012)
Edited by Marie Gillespie and Alban Webb

Making Culture, Changing Society (2013)
Tony Bennett

Rio de Janeiro: Urban Life through the Eyes of the City (forthcoming)
Beatriz Jaguaribe

Interdisciplinarity: Reconfigurations of the Social and Natural Sciences (forthcoming)
Edited by Andrew Barry and Georgina Born

Devising Consumption: Cultural Economies of Insurance, Credit and Spending (forthcoming)
Liz Mcfall

Unbecoming Things: Mutable Objects and the Politics of Waste (forthcoming)
By Nicky Gregson and Mike Crang

Centre for Research on
Socio-Cultural Change

Making Culture, Changing Society

Tony Bennett

Routledge
Taylor & Francis Group

LONDON AND NEW YORK

First published 2013
by Routledge
2 Park Square, Milton Park, Abingdon, Oxon OX14 4RN

Simultaneously published in the USA and Canada
by Routledge
711 Third Avenue, New York, NY 10017

Routledge is an imprint of the Taylor & Francis Group, an informa business

© 2013 Tony Bennett

British Library Cataloguing in Publication Data
A catalogue record for this book is available from the British Library

Library of Congress Cataloging in Publication Data
Bennett, Tony, 1947-
  Making culture, changing society / Tony Bennett.
    p. cm. -- (Culture, economy and the social)
  Includes bibliographical references and index.
  1. Culture. 2. Social evolution. 3. Social change. I. Title.
  HM621.B453 2013
  303.4--dc23
                  2012034139

ISBN: 978-0-415-68884-0 (hbk)
ISBN: 978-0-203-33232-0 (ebk)

Typeset in Times New Roman
by Taylor & Francis Books

MIX
Paper from
responsible sources
FSC
www.fsc.org   FSC® C013604

Printed and bound by CPI Group (UK) Ltd, Croydon, CR0 4YY

To Sue – again and always

Contents

Figures

Acknowledgements

Those who deserve our thanks the most usually end up being mentioned last in authors' acknowledgements. Not this time: the person I need to thank first is my wife and partner, Sue. This has always been so but, this time, more than ever for holding together so many aspects of our lives over the long haul of moving back from Britain to Australia. This book was conceived and begun on one side of that process, in 2008, but only finished several years later on a different side of the world. First and foremost, then, my love and thanks to Sue for keeping our lives afloat. There would have been no book otherwise.

This move also involved a significant change of institutional and intellectual contexts: from the Open University to the University of Western Sydney, and from the former's involvement, with the University of Manchester, in CRESC (the ESRC Centre for Research on Socio-cultural Change) to the latter's Centre for Cultural Research, now the Institute for Culture and Society.

The conception of this book owes a good deal to the intellectual stimulation I derived from working in CRESC with many colleagues across its two partner universities. This was particularly true of the members of the research theme on 'Culture, governance and citizenship: the formation and transformations of liberal government'. I owe a good deal to Patrick Joyce, with whom I convened this theme. Patrick's influence is evident in many of the historical aspects of my argument. And I owe a good deal too to Francis Dodsworth, from whom I have learned much about eighteenth-century debates on liberty which, although I do not draw on them directly here, helped me in understanding the later, and different disposition, of the relations between government and freedom in nineteenth-century forms of liberal government. I am also indebted to Patrick and Francis for introducing me to the the historians whose work on governmentality, histories of freedom and material infrastructures informs many of the arguments in what follows.

I am similarly indebted to the other members of this theme who shared – and share – my sense of the importance of museums as key loci for the concerns of contemporary cultural theory. Working with Helen Rees-Leahy and Sharon MacDonald, both from the University of Manchester, proved very helpful in this regard; I learned a huge amount from Helen's extensive knowledge of the history and practice of art galleries, and from Sharon's work on difficult heritage that, alongside her work on science museums, have constituted such a distinctive contribution to the field. I learned much, too, from Grahame Thompson's work on networks and markets and his general capacity to provoke new insights, and from Sophie Watson's work on 'city materialities'. I owe Sophie, though, a good deal more for her consistent friendship and support throughout the years we worked together at the Open University, and since.

I owe a good deal, too, to Nik Rose whose work on the relations between government and freedom – published before Foucault's formulations on this couplet were widely known – has influenced significant aspects of my argument. I was, in this regard, especially pleased to have the opportunity of working closely with Nik in the ESRC seminar series that he, Patrick Joyce, Francis Dodsworth and I convened on the theme 'Government and Freedom: Histories and Prospects' in 2008–9 (Award No. RES-451-26-0390). I acknowledge my indebtedness to the many contributors to this series whose presentations helped me in framing my approach to the relations between culture, government and freedom in this book. My particular thanks to Mariana Valverde and James Hay in this regard.

Still in CRESC territory, and with regard to those aspects of this book that engage with the work of Pierre Bourdieu, I should like to acknowledge the debt I owe to Mike Savage, Elizabeth Silva and Alan Warde with whom I worked closely in the ESRC project on 'Cultural Capital and Social Exclusion' (Award No. R000239801). I draw on aspects of this work in Chapters 7 and 9. I owe Mike a lot more in view of the long experience we had in working together – alongside Karel Williams and, later, Marie Gillespie and Penny Harvey – in directing CRESC through its initial years. Apart from his great energy and openness, Mike's interest in the social life of research methods converged in many interesting and helpful ways with my interest in the knowledge practices of collecting institutions. This was also true of Evelyn Ruppert's work in the same area – I draw on this directly in Chapter 5 – and, more generally, of John Law's innovative work at the interfaces of actor-network theory and social science methods.

There are other debts to record to CRESC colleagues. I should like, first, to acknowledge how much I have benefited from the experience of co-editing the series in which this book appears with Penny Harvey and Kevin Hetherington. I have especially valued the insight that working with Penny has given me into the new work and debates that are currently reshaping anthropology. I have equally valued working with Kevin. We share many interests, particularly in the history and theory of museums, but from angles of vision that are slanted slightly differently. I am, therefore, especially grateful that Kevin was able to read the manuscript for this book. I found his detailed and probing comments invaluable and, though I doubt that I have done them full justice, am particularly grateful for the pains he went to in this regard. My final CRESC acknowledgement is to Liz McFall and Mike Pryke with whom I co-edited the *Journal of Cultural Economy* for its first five years. I have, through their work and the networks they are connected to, become familiar with many literatures I would not otherwise have had access too. Although I do not pay any specific attention to cultural economy debates in what follows, the literatures that define this field have significantly informed my approach to questions concerning the relations between culture and the social. My continuing involvement with both this journal and this book series has been facilitated by my appointment as a visiting professor at the Open University. I acknowledge my appreciation of this support and the good offices of both Tim Jordan and Kath Woodward in securing it.

I turn now to acknowledge my appreciation of the contribution that Australian institutions and colleagues have made in supporting the work that has gone into this book. I am particularly grateful, first, to the University of Melbourne for offering me an appointment as a visiting professor in its School of Culture and Communications, and to John Frow and Stuart McIntyre for securing my appointment to this position. I thank John, too, for the continuing value of intellectual exchange over a long period, and particularly for his supportive comments on the manuscript for this book. I also thank Chris Healy with whom I worked closely during my biannual visits to Melbourne, and particularly for the experience of working with him in the CRESC/University of Melbourne workshop on 'Assembling Culture' that we jointly convened.

My second set of thanks goes to my colleagues at the University of Western Sydney. I am most grateful, first, to Wayne McKenna for interesting me in joining UWS and for his role in making my move back to Australia such a welcoming experience. My thanks, too, go to David Rowe, the Director of the Centre for Cultural Research at the time, for his support in facilitating my transition to UWS and to his successor in this role, Brett Neilson. A more particular note of thanks is owed to Ien Ang for her friendship and generosity in welcoming me to UWS, and for the pleasure of working with her in transforming the Centre for Cultural Research into the Institute for Culture and Society. The environment that Ien and her colleagues have created at UWS over many years has proved invaluable in enabling me to see this project through. So too has been the research support that I have received, especially through the research assistance of Michelle Kelly – whose extraordinary thoroughness and efficiency has been of immense value – and the equally impressive support I received in the performance of my other duties from Reena Dobson.

Moving to UWS has also resulted in new intellectual collaborations that have had a direct bearing on many parts of this book. I acknowledge first here my debt to Greg Noble and Megan Watkins for their shared interest in the subject of habit. Chapters 8 and 9 draw considerably on the various exchanges we have had on this subject as well as on the experience of working with them, and with Francis Dodsworth and Mary Poovey from NYU, in jointly organising a research workshop on the topic of habit and governance. My thanks, too, to the other participants in this workshop, and particularly to Barry Hindess whose work has helped to shape my own in many ways. I am grateful also to Fiona Cameron who shares my interest in the relations between museums, governmentality and assemblage theory, but brings to these a fresh perspective that has been of considerable benefit to me. I acknowledge too the value of working with Fiona in applying for, and subsequently developing, the Australian Research Council project on 'Museum, Field, Metropolis, Colony: Museums and Social Governance' (Award No. DP110103776). This has in its turn generated new collaborations with my UWS colleague Ben Dibley and with a distinguished team of international collaborators whose expertise on the relations between museums and anthropology I have drawn on at various point in the book. My thanks, then, to Nélia Dias from the Department of Anthropology in the University of Lisbon, to Rodney Harrison from, initially, the Open University but now from UCL's Institute of Archaeology, to Ira Jacknis from the Phoebe A. Hearst Museum of Anthropology at the University of California, Berkeley, and to Conal McCarthy from the Victoria University of Wellington in New Zealand. My particular thanks in this regard to Nélia for reading and commenting on Chapter 5. And a particular note of thanks to Kay Anderson whose work has helped stimulate my interest in post-humanist debates and whose comments on an early version of Chapter 8 proved of real assistance.

There is no better learning experience than that provided by the opportunity to present and discuss one's work with other scholars, researchers and students. Many of the arguments, contentions and perspectives informing the following pages have benefited immensely from exchanges of this kind. I am, then, grateful both to those who have invited me to present work at such occasions, and to the participants for their comments. Early versions of different parts of Chapter 1 were first presented at the 2010 annual conference of the UK's Media and Cultural Studies Association conference at the London School of Economics, and at the 2010 Cultural Crossroads conference at Lingnan University in Hong Kong. My thanks are due to Sonia Livingstone at LSE, and to Meaghan Morris, Stephen Chan and their co-organisers at Lingnan for providing me with these opportunities.

The arguments informing Chapter 2 were first presented at the inaugural conference of the Institute for Cultural Theory, University of Manchester, and subsequently at the Kaifeng Forum on Literary Theory and Cultural Studies held at Hunan University, and at CRESC's 'Culture and Social Change: Disciplinary Exchanges' conference in 2005 where it comprised part of a strand of related papers convened by Patrick Joyce and myself. A selection of these papers was published as a special issue of the journal *Cultural Studies* (21, 4–5, 2007) on 'Liberalisms, government, culture'. This chapter also draws on work presented at the international symposium on 'Cultural Studies and Institution' held at Lingnan University in 2007 and at the 2007 annual conference of the Swedish Cultural Studies Association held in Norrköping. My thanks to Meaghan Morris again, and to Johan Fornäs for inviting me to take part in these events.

Chapter 3 derives, first, from a keynote address at 'The Rebirth of the Museum' conference organised by the University of Melbourne and the

Melbourne Museum in 2004. I am grateful to Chris Healy for inviting me to take part in this conference and to the conference delegates – particularly Andrea Witcomb and James Clifford – for their critical feedback. These initial arguments were later developed in a public lecture presented at the Vancouver Art Gallery and University of British Columbia in 2005, and in a paper presented at 'The Global Future of Local Art Museums' symposium at the University of Vienna's Internationales Forschungszentrum Kulturwissenschaften in 2006. I am grateful to Hans Belting for inviting me to take part in this symposium and for his helpful observations on the paper I presented.

The arguments eventually brought together in Chapter 4 had their first dress rehearsal in an address presented at the conference 'Nature Behind Glass: Historical and Theoretical Perspectives on Natural Science Collections' arranged by the Museums and Galleries History Group and the Society for the History of Natural History and held at the University of Manchester in 2007. My thanks to Sam Alberti for arranging my participation in this conference. These arguments were subsequently developed at a 2008 CRESC workshop on 'Museum Knowledges and the Transformation of the Social'. My thanks to Helen Rees-Leahy and Sharon MacDonald for their roles in co-organising this workshop. Chapter 5 draws on work first presented at the 'Assembling Cultures' workshop organised by CRESC, the Australian Cultural Researchers Network, and the Ian Potter Foundation, and held at the University of Melbourne in 2010. Special thanks to all the participants in this event, subsequently published as special issue of the *Journal of Cultural Economy* 2009, 2 (1–2). It also benefits from the discussions I was involved in during the seminar 'Reassembling the Collection: Indigenous Agency and Ethnographic Collections' held at the School of Advanced Research, Santa Fe, New Mexico in 2010. I am grateful to Rodney Harrison for arranging my participation in this exceptionally rich research seminar.

Chapter 6 draws on work presented at the 'Literature and Liberation' conference held in celebration of the work of Graham Martin that was organised by the Institute of English Studies, University of London, the Raymond Williams Society and the Open University; at the 'Art in the Lifeworld' conference held at Ballymun Community Arts Centre, Dublin in 2008; and as a public lecture for the Institute for Advanced Studies in the Humanities and Social Sciences, Nanjing University. I am grateful to Professor Wang Jie, the Institute's director, for inviting me to present this lecture and for his generous hospitality during my visit to Nanjing. Chapter 7 was first presented at the conference 'Interpretation, Theory & the Encounter' held at Tate Britain on 9 July 2010. I am grateful to the conference organisers, Sylvia Lahav and Victoria Walsh, for inviting me to take part in this conference, and for their inclusion of the paper in a special issue of *Tate Papers*. I am also grateful to Jennifer Mundy for her invaluable assistance in finalising this text.

Earlier versions of Chapter 8 were presented as a Distinguished Lecture for the Council on Culture and Media at New York University, and at 'Sustaining Cultures', the 2007 annual conference of the Cultural Studies Association of

Australasia. My thanks to Susan Luckman for arranging this latter invitation. A more developed version was presented at the symposium on 'Liberal Subjects: The Politics of Social and Cultural History since the 1980s', convened, at the University of Manchester, by Simon Gunn and James Vernon to commemorate the work of Patrick Joyce. I am greatly indebted to Simon and James for their editorial comments on this paper. Chapter 9, finally, draws on a public lecture presented at the Department of Comparative Literature, at Nanjing University, and at the Department of Chinese, Fudan University, Shanghai. It also draws on a paper presented at the 'Thirty Years after Distinction' conference that was held in Paris in 2010, organised by the Observatory for Social Change in Sciences-Po, the Centre of European Sociology and the Laboratory of Quantitative Sociology at INSEE (National Institute of Statistics). I thank Philippe Coulangeon and Julien Duval for the opportunity to take part in this landmark event.

Four of the chapters have been published before as either journal articles or as chapters in edited collections. I have, however, edited these, and in some cases quite significantly, so as to delete any redundancies that might arise from bringing them together with other chapters; to make connections with other parts of the book in order to clarify their place within the argument of the whole; to amplify particular arguments and, occasionally, to correct points of detail. This is true of Chapter 3, which is derived from Bennett (2005a); of Chapter 4, derived from Bennett (2010); of Chapter 7, derived from Bennett (2011a); and of Chapter 8, derived from Bennett (2011). The References at the end of the book provide the full details for these sources. The remaining chapters have been written specially for this book although, in some cases, they draw on selective aspects of earlier publications. Chapter 2 draws on Bennett (2006, 2007, 2007a, and 2010b), Chapters 5 and 6 on aspects of Bennett (2009) and Bennett (2008/9) respectively, and Chapter 9 on elements on Bennett (2005 and 2011b). I appreciate the support of the co-authors and publishers of *Culture, Class, Distinction* (Bennett et al., 2009) in agreeing to my reproducing illustrations from that book as Figures 7.1 and 7.2.

Note on the text

I have tried not to clutter the text with too much detail so far as references are concerned. When I first refer to a source in any chapter, I give the author's name, date of publication and page reference, but only the page reference in any immediately following references to the same source. Where italics are used within quotations this is faithful to the original. Translations from French sources are my own.

Introduction

This book both continues and departs from lines of inquiry I have pursued in earlier work concerned with the relationships between culture and practices of governance. I first engaged with these questions in *Culture: A Reformer's Science* (Bennett 1998), partly with a view to defining a place for cultural policy studies within the broader field of cultural studies. This required, I argued, that the concept of culture be rethought so that such concerns might be presented as integral to the field rather than being grudgingly admitted as somewhat extraneous, largely pragmatic, additions to it. I drew for this purpose on Foucault's concept of governmentality in view of the critical leverage it provided in relation to those uses of the concept of culture – mainly those current at the time within Anglophone cultural studies – in which culture and government were understood as contraries. I took a similar tack in *Pasts Beyond Memory* (Bennett 2004), but one which placed a greater emphasis on the role of knowledge practices by examining how the museological deployment of post-Darwinian developments in the life sciences, archaeology, and anthropology informed the changing forms of late-nineteenth-century liberal government in Australia, Britain and the United States. I sought, in doing so, to bring the perspectives of governmentality theory into a productive relationship with those of science studies, particularly as represented by Bruno Latour's work on centres of calculation. This involved a consideration of the respects in which museums operated as critical switch points across different networks: centres of calculation that, in ordering the materials they accumulated from diverse points of collection, produced new entities that they then relayed back out to the world as resources for shaping conduct via their deployment in different networks: the capillary networks of public schooling, those of public spheres or the apparatuses of colonial or social administration.

Making Culture, Changing Society connects these interests in the relations between governmentality theory and knowledge practices to the perspectives of assemblage theory and the more general 'material turn' that is currently in evidence across the human and social sciences in order to explore their implications for how we might now best use and interpret the concept of culture. I am, though, less concerned to argue for an interpretation of the concept that would find its primary *rendez-vous* within cultural studies than *after culture?*

Culture as historically & institutionally-embedded

I am to explore whether and, if so, on what conditions the concept is capable of identifying a meaningfully specific and coherently integrated set of concerns. I say 'if' as the continuing value of, and need for, the concept are no longer matters that can be taken for granted. Few of the theoretical traditions that I draw on in this book have a great deal to say about the concept of culture. They have also, between them, rung the death knell for the 'cultural turn' as, in essence, a linguistic modelling of the social. This 'turn' has proved productive as a counter to the tendency to seek the 'truth' of culture in a set of underlying economic and social relations that (in however qualified a way) characterised the Marxist debates of the 1960s and 1970s. Its chief limitation, though, is that of sharing a similar problem space in which the analysis of the relations between culture and society come to pivot on the decipherment of a general set of relations between representations and the real.

It is for these reasons that I ask, as my starting question, whether we might now need to relocate the concerns that have previously been addressed under the heading of culture on an analytical terrain that is 'after culture'. Although I answer this question in the negative, my reasons for posing it are not solely rhetorical. It is only by probing its limits that the concept of culture can be brought into a productive alignment with a theoretical landscape that is now significantly at odds with many of the intellectual coordinates that have shaped its history. My wager, in what follows, that this might be done involves four main lines of argument.

The first is a historical one to the effect that, rather than being equated with the realm of the symbolic and thus be interpreted as a component of all societies, culture is better regarded as a historically specific set of knowledge practices. I argue that these practices are the product of specific forms of expertise performed in particular institutional settings whose interactions comprise what I call a 'culture complex': that is, an ensemble of institutionally embedded knowledge practices that are entangled within, and act on, economic and social relations in varied ways. I work through these arguments by drawing on Michel Foucault's understanding of the operations of knowledge practices as materially entangled components of *dispositifs,* and on Gilles Deleuze's conception of the diagram as a means of understanding the dispersed, decentred intersections of power relations through which the social is constituted in the interactions between such *dispositifs.*

These contentions are closely related to my second main line of argument: that the mechanisms through which the culture complex is connected to, acts on and shapes the social proceed not through some general mechanism of cultural constructivism but through a distinctive ontological politics of culture. The conception of ontological politics that I draw on here had its original provenance in the field of science studies where – as in Annemarie Mol's (2002) work on the medical sciences, for example – it refers to the capacity of scientific practices to produce and perform new realities through the alignments of the relations between both human and non-human actors that they effect. It has now, however, acquired a much broader currency, particularly in

[handwritten: foucauldian arg.]

[handwritten: "architects of the person"]

its application to the performative capacities of the social sciences. There have been two main fields of application here: the 'cultural economy' tradition that, taking its primary cue from the work of Michel Callon and his collaborators (Callon, Millo and Muniesa 2007), has demonstrated the role that the various sub-disciplines of economics play as forces intervening in the makeup of their objects of study (markets), and work on 'the social life of methods' that has investigated how social science methods produce and perform new realities. There has, though, been little attempt to apply these perspectives systematically to the social role of cultural knowledges: art history, aesthetics, archaeology, literary studies, anthropology and so on. This is the endeavour that I seek to contribute to here. I do so by considering how far Bruno Latour's contentions regarding the ways in which scientific practices produce new collectivities of actors (human and non-human) whose agential capacities arise from the orchestration of their inter-relations can illuminate the social lives of cultural knowledges. While I indicate how the concepts and arguments I propose are more generally applicable, my historical focus is, for the greater part, on the nineteenth and early twentieth centuries, while the knowledge practices I devote most time to are aesthetics and anthropology.

The reasons for this have to do principally with the third line of argument that runs through the book. This concerns the respects in which the knowledges and institutions comprising the culture complex form parts of a historically distinctive regime of government, which, in ordering freedom, also distributes it differentially through the social body, as well as across the relations between different types of society. The historical and theoretical coordinates for this argument are derived from Michel Foucault's general formulations of liberal govement and, in particular, from the arguments he put forward in *The Birth of Biopolitics* (Foucault, 2008) regarding how liberal forms of government work by and through the distinctive kinds of freedom that they produce. Foucault is clear that these freedoms are made up through the operation of distinctive regimes of veridification or truth practices that, among other things, regulate the (unequal) distribution of the capacities for freedom that they produce and enjoin across different populations. How have the knowledge practices of anthropology and aesthetics been tangled up in the exercise of these forms of liberal government? How have they been implicated in the makeup and distribution of different kinds of freedom both within metropolitan powers and across the relations between these and colonial contexts?

A part of my answer to these questions depends on my fourth and final general line of argument. This concerns the role that cultural knowledge practices play in producing what I call different 'architectures of the person' as their routes into the direction of conduct. Again, the provenance for my argument here is broadly Foucauldian, drawing on the ways in which Foucault's work on technologies and practices of the self proposes an internal partitioning of the self in order to provide the differentiated spaces across which the action of self on self can take place. However, it also looks, as did Foucault, to the work of Pierre Hadot (1995), whose analysis of the role of spiritual exercises in the conduct of philosophy as a distinct 'way of life' has had a broader influence on the distinctive forms of personhood associated with different kinds of scientific and intellectual practice. Lorraine Daston's and Peter Galiman's (2010) account of the different organisations of the scientific self associated with the transition from Enlightenment conceptions of 'truth to nature' to mid-nineteenth-century conceptions of objectivity is a case in point. My interest in 'architectures of the person' is focused on how intellectual practices in the social and human sciences lay out differentiated components of personhood and orchestrate their relations to one another in ways that provide intellectual authorities with conduits into the regulation of conduct. They provide points of leverage through which distinctive forms of cultural expertise access and enable specific capacities; and they do so, I shall argue, in ways that differentiate populations depending on the degree to which the architectures of personhood attributed to them afford, or deny, a capacity for freedom. By pursuing these concerns in relation to the role that habit is accorded within different architectures of the person, I explore how the distribution of a capacity for freedom is articulated across a liminal zone between the human and non-human.

Part 1

Culture: veridical, material and compositional perspectives

1   After culture?

I am by no means the first to ask whether the concept of culture might have outlived its usefulness. Its value has been queried from a number of different perspectives. Adam Kuper,. noting the multiplication of its uses, recommends abandoning the concept in favour of a range of more specific terms: belief, art, custom, tradition (Kuper 1999: x). This reflects a widespread tendency within contemporary anthropology to eschew the term, not least because of its relations to the entangled histories of anthropology and colonialism. Niklas Luhmann takes a similar tack, suggesting that, from the point of view of understanding the dynamics of the modern art system, culture has proved to be 'one of the most detrimental concepts ever invented' (Luhmann 2000: 247). Richie Nimmo (2010), stressing the entanglements between the concept and the species-centrism of Western humanism, similarly urges that we should now put it to one side if we are to respond adequately to the ecological imperative of better understanding the relations between human and non-human actors. James Clifford's assessment is more equivocal. While conceding that the concept of culture might have served its time, he also urges that whatever concept finally transcends it should preserve its 'differential and relativist functions' while avoiding 'the positing of cosmopolitan essentialisms and human common denominators' (Clifford 1988: 274–5).

I take a somewhat different approach. It is perfectly clear that there are now ways of engaging analytically with the practices that have conventionally been brought under the heading of culture that do not have much space for a concept of culture as such. This is true of the varied branches of assemblage theory and actor-network theory. These have undoubtedly stimulated new approaches to the social entanglements of cultural practices.[1] Yet none of these has had much to say specifically about culture as a general concept. More to the point, these traditions have been significant points of reference for many of those who have suggested that we jettison the concept entirely. This is true of those post-humanist theorists who, via Deleuze, claim an affiliation to what Jane Bennett calls the 'vital materiality' derived from Henri Bergson's work (Bennett 2010: v11). I draw strategically on these traditions in what follows. I do so, though, with a view to suggesting ways in which they might contribute to a critically renovated concept of culture that, in limiting its application to a specific set of historical processes, will sharpen its analytical purchase.

I pursue these concerns by exploring the interfaces between intellectual traditions in which the concept of culture has played a pivotal role and those more recent traditions, briefly identified above, which have been either indifferent or hostile to it. So far as the former are concerned, I focus mainly on Anglophone cultural studies and on cultural sociology as represented by Pierre Bourdieu and the work that has developed in his wake. There are significant differences between these traditions.[2] From the perspective of my concerns here, however, both describe a critical orbit around Kantian conceptions of culture, constantly seeking to pull away from it but without ever escaping its gravitational pull.[3] That pull is exercised through different way stations: Matthew Arnold and English in the case of cultural studies, Émile Durkheim and sociology in the case of Bourdieu. Recent work has considerably weakened the hold that Kant's work has exercised over the social and cultural sciences,[4] and I will draw on this to show how key aspects of both these traditions remain in thrall to the conception of culture as a process of collective human fulfilment that Kant proposed. For it is this aspect of the Kantian legacy that is now most in question.

There are, of course, already vast literatures exploring the relations between these cultural and 'a-cultural' analytical territories. Deleuze has long figured as a force for critical renovation within cultural studies and cultural sociology,[5] yet often in ways that align his concerns with long-standing vocabularies of culture and society. Larry Grossberg, for example, argues a case for forging strong connections between Deleuze's concepts and those of cultural studies as represented by Raymond Williams and Stuart Hall: between Hall's concept of conjuncture and Deleuze's concepts of milieu, territory and diagram, for example (Grossberg 2010). In sociology, by contrast, post-Deleuzeian developments in assemblage and actor-network theory have attempted to rethink the social in ways that will detach it from its Durkheimian–Bourdieusian lineage.[6] My interests tend more in this second direction in the sense that I am less concerned to explore areas of possible *rapprochement* than those of dissonance between such post-Deleuzian traditions and those formulations of the relations between 'culture' and 'society' associated with cultural studies and cultural sociology.

The place from which I conduct this work is, broadly speaking, that provided by Michel Foucault's perspective of governmentality. This has been drawn on in what is now a quite extensive literature to examine how culture has come to constitute, in George Yúdice's telling phrase, an expedient resource for the governance of contemporary populations (Yúdice 2003). I develop this aspect of Foucault's work by interpreting what little Foucault had to say directly about the concept of culture in the light of his more general methodological precepts. These will provide two key building blocks for the approaches to the interpretation and analysis of culture I shall propose. The first of these derives from what Thomas Osborne calls the '"veridical" twist' that Foucault brings to the concept of culture (Osborne 2008: 70). Insofar as it comprises a set of resources involved in the governance of populations, culture operates through the distinctive regimes of truth and forms of expertise that it instantiates. The second building block derives from the methods Foucault deploys in historicising objects of analysis that are commonly taken to be universal. By aligning these two perspectives I shall suggest that culture is best interpreted as a historically bounded set of truth practices that are implicated in regulating the 'conduct of conduct' in specific ways through their operations as parts of assemblages that are differentiated from, and ordered in specific relations to, the social and the economy. I shall, though, want to part company with Foucault so far as his accounts of Kant and the Enlightenment, and their implications for his understanding of the relations between culture and critique, are concerned. In his susceptibility to the legacy of post-Kantian aesthetics, Foucault sometimes remained caught within the 'machinery of culture' rather than providing a critical purchase on that machinery.

These are matters that I pursue in greater depth in due course. My more immediate concern is to amplify the veridical perspective I derive from Foucault and to identify its relationship to his procedures for historicising objects of analysis. I shall then return to elaborate more fully why the concept of culture now seems to be increasingly 'unhinged', in the sense of being unable to meet the conditions required to secure its coherence, and to outline the ways in which I propose to address these conditions.

Historicising culture

I have already noted that the concept of culture does not figure prominently in Foucault's theoretical vocabulary. There is, indeed, only one place where he accords the concept serious attention, and even then he is wary of it. While 'having trouble with the word and putting it in inverted commas', Foucault allows, in *The Hermeneutics of the Subject*, that it might be possible to speak of 'culture' provided that four conditions are met. First, there has to be a set of values 'with a minimum degree of coordination, subordination and hierarchy' and, second, these values have to be 'given both as universal but also as only accessible to a few' so as to produce 'a mechanism of selection and exclusion'. The third condition is that 'a number of precise and regular forms of conduct are necessary for individuals to be able to reach these values', and the fourth – the 'veridical twist' that Osborne identifies – is that the techniques for acquiring those values have to be taught, transmitted and validated as parts of the operation of a 'field of knowledge' (Foucault 2005: 179).

This is quite an extensive definition capable of spanning, for example, the exclusionary logic of the 'spiritual exercises' of Greek, Hellenistic, Roman and medieval Christian philosophies that Foucault addresses in this lecture series and, later, in *The Care of the Self* (Foucault 1988). It can also encompass Max Weber's concern with the relations between Calvinism and the 'spirit of capitalism' (Weber 2001) and Pierre Bourdieu's account of the role played by institutions of cultural legitimation in validating and selectively transmitting those techniques of appreciation that permit only selective access to the 'universal' values of the Kantian aesthetic (Bourdieu 1984). It is, however, the role of culture in distributing the capacity for certain forms of self-governance unequally across different sections of the population, rather than its role in producing a specific ethic of economic conduct or in organising class divisions, that gives Foucault's concept of culture its analytical and political coherence. For it is the capacity for self-governance that qualifies those who (claim to) possess it to govern those whom they judge to lack it.

Of course, these concerns come together at crucial junctures, and I shall explore these at appropriate points in my discussion. Nonetheless, Foucault's point of entry into them is distinct insofar as the relations between governors and governed do not turn on the single axis of economic divisions but, depending on the circumstances, may revolve primarily around divisions of age, gender, sexuality, coloniser/colonised or status (freeman/slave, for example). This has considerable advantages over those concepts of culture (like Bourdieu's), which, having identified its role in terms of a primary articulation across economic or class relations, are then obliged to account for its operation across other relations – of gender or ethnicity, for example – in accordance with the logic of the 'sociological supplement' in which other considerations are tacked on as added 'variables'. Foucault's approach does not suppose, look for or require a primary axis of differentiation for the exercise of power. His formulations also have advantages over Jacques Rancière's assessment of the role that philosophical and aesthetic practices play in organising distinctions between governors and the governed (Rancière 2003). For they do not, as does Rancière, limit the operation of those practices to the social division between occupations – a limitation whose significance I consider in Chapter 7. Nor do they limit the sphere of culture to that of the aesthetic. They clearly include it, but only if understood as one amongst many forms of cultural expertise.

I do not, though, think that Foucault's brief comments on the subject constitute a ready-made peg on which to hang a theory of culture. To the contrary, while a useful point of departure for such a theory, they also stand in the way of its development. They do so, moreover, precisely because of what might, at first sight, seem to be their chief advantage. This consists in their pliability in seemingly offering the basis for a general account of how knowledge practices connect with social processes that might be aligned with post-Durkheimian interpretations of culture as a set of trans-historical and trans-societal processes focused on the role played by the symbolic in the organisation of social life. Yet this would be to transform Foucault's work into a sociology, which it is surely not. To the contrary, we can hear in Foucault's hesitancy to embrace the concept of culture and, equally, in his failure to ever engage with Bourdieu's work,[7] his reluctance to get tangled up in the Durkheim–Bourdieu lineage that, at the time he was writing, defined the intellectual trajectory of French cultural sociology.[8] For it is precisely the universalism of the Durkheimian conception of the symbolic that runs against the grain of what I take to be the more valuable legacy of Foucault's work: the radical historicality of his methodological perspective.

He insists on this quite trenchantly in his course of lectures on the birth of biopolitics where he takes issue with the kind of procedure informing his later – and as we have seen, tentative – definition of culture. Rather than start from universals like the state, the people, subjects, sovereignty or civil society as 'an obligatory grid of intelligibility' for the analysis of concrete governmental practices, Foucault argues, analysis should start with such practices and, 'as it were, pass these universals through the grid of these practices' (Foucault, 2008: 3) rather than vice versa. Distinguishing his procedures from those of sociology, history and political philosophy, Foucault takes issue with the ways in which the relations between historically concrete governmental practices and conceptual universals are usually understood within these disciplines. Bringing their methods together under the heading of 'historicism', which he interprets as an intellectual practice that takes such universals as a given and sets out to see how history variably inflects or alters them, he urges an inversion of the methodological orientations this involves:

> Historicism starts from the universal and, as it were, puts it through the grinder of history. My problem is exactly the opposite. I start from the theoretical and methodological decision that consists in saying: Let's suppose that universals do not exist. And I then put the question to history and historians: How can you write history if you do not accept a priori the existence of things like the state, society, the sovereign and subjects?
>
> (3)

Foucault gives an example of how to answer this question when, in the final lecture in the series, he disputes the intelligibility of the political–philosophical conception of civil society as 'a primary and immediate reality' or as 'an historical-natural given' that always functions 'in some way as both the foundation of and source of opposition to the state or political institutions' (297). It is, to the contrary, Foucault argues, a product of modern governmental technologies. This does not mean that its ontological status is in any way diminished; it is not, Foucault insists, a mere construct. Its status is rather that of a 'transactional reality': that is, a reality that, far from constituting at any particular historical moment the variable form of a set of institutions and practices that are external to the state, is a product of those specifically self-limiting forms of modern liberal government that produce civil society as a historically novel interface in the relations between governors and governed.

Culture is not among the universals Foucault suggests should be historicised in this way. My point, though, is that it should be. For its emergence, as a concept and as a set of historically operative realities, is coeval with those that he does name – the state, the people, subjects, sovereignty, civil society – and, is moreover, clearly tangled up with these as well as with the emergence of another set of 'conceptual universals': nature, economy and society. I propose, then, to treat it in the same way: that is as a historically specific 'transactional reality' that has its locus in specific governmental practices and technologies and which has to be considered in terms of its relations to a similar historical specification of these other 'universals' if its *modus operandi*, spheres of action and effects are to be properly understood.

It is from this perspective that I want to look now at the key difficulties that currently beset the concept of culture. To dwell on these, though, might seem a little paradoxical at a time when culture is increasingly invoked as a key connecting term across the social sciences and humanities while also figuring increasingly prominently in public, policy and political discourses: the endless urging of the need for 'cultural' solutions to be sought where social divisions seem intractable, the importance of the 'cultural economy' and so on. The difficulties I am concerned with, however, operate at a different level and derive, ultimately, from culture's relationship to the other 'conceptual universals' it has been most closely connected to. We are now accustomed to the project of a 'sociology without society' (Urry 2000) as the sociological concept of society has increasingly been replaced by variant formulations of the social as a historical effect of regulatory and governmental practices (Joyce 2002) or as the outcome of specific processes of assembling (Latour 2005). The economy is similarly increasingly understood as a product of relatively recent processes of 'economisation' (Caliskan and Callon 2009, 2010) while Bruno Latour has identified the challenge of what it might mean to govern the world 'now that Nature as an organising concept (or, rather, conceit) is gone' (Latour 2010: 479). What are we now to make of the concept of culture when the key terms in relation to which its distinguishing qualities have typically been defined are increasingly interpreted as the outcomes of historical processes of assembling that have called their earlier universal status into question?

There are, viewed in this light, three crucial suppositions that have underlain the concept of culture that these developments call into question by dismantling the intellectual coordinates that have informed its definition. The first consists in the supposition that culture might be identified as a specific realm of practices, which is distinct from both the social and the economy and which acts on these in terms of the properties that distinguish it from them. The second is that specific forms of cultural practice might be distinguished from others to provide the basis for a distinctive cultural politics that depends on the production of particular kinds of free and self-conscious subjectivities. The third supposition is that culture is to be defined in terms that restrict it to – indeed, are constitutive of – a uniquely human set of practices, thus distinguishing it from nature and, more generally, effecting a division between human and non-human actors. I shall, then, look more closely at each of these suppositions. In doing so I also indicate how the veridical and historical perspectives that I have briefly outlined might serve to refashion a more viable interpretation of the concept of culture.

Culture in question (i): after representation

The first assumption, to restate it briefly, holds that culture might be identified in terms of properties that distinguish it from economic and social practices. This is the assumption of the tradition of French cultural sociology, running from Durkheim through to Bourdieu, whose central concern is with the role of the symbolic in the organisation of social life.[9] It is also the assumption of the Anglophone traditions of cultural studies in which culture is defined as the realm of meaning-making practices to be considered in terms of their conditioning by, and consequences for, the conduct of economic and social practices. It is finally, and more generally, the logic underlying the 'cultural turn' that has drawn on both of these traditions to propose an active role for culture in the construction of social and economic life. This role is exercised through the influence that cultural representations exert on how social and economic agents view and interpret their own actions, identities and relations to one another.

These variant formulations are called into question by the challenge that has been posted by 'post-representational' perspectives to the very enterprise of defining culture as a reality of a particular type (made up of the symbolic, meanings, representations) that is distinct from economic or social realities. Following in the wake of Foucault's work, particularly his concept of the *dispositif*, and seeking a counter-heritage in the work of Gabriel Tarde, the traditions of actor-network and assemblage theory deploy what David Toews (2009) has usefully called a 'compositional perspective'.[10] In place of the great founding separations of the nineteenth-century social sciences between culture, society and economy, this focuses on the more historically specific 'gatherings' of varied elements (textual, technological, human, non-human) into provisional associations with one another that traverse such great divides. As Michel Callon puts it:

> There isn't a reality on the one hand, and a re-presentation of that reality on the other. Rather, there are chains of translation. Chains of translation of varying lengths. And varying kinds. Chains which link things to texts, texts to things, and things to people. And so on.
>
> (Callon and Law 1995: 501)

In exploring the implications of these perspectives, my contention will be that they provide a basis for a more concretely delimited specification of culture provided that it is understood as the historically mutable and contingent product of such processes of gathering. It is a linked series of assemblages that bring together an array of heterogeneous elements and organise them into distinctive compositional configurations from which its distinguishing capacities derive. The 'culturalness' of culture, on this conception, is not given by the general properties of the symbolic or the logic of representation, but consists rather in the nature of the gatherings that are produced by the ordering of the relations between the elements that constitute it.

This shifts the level of analytical specification to the operations of what I shall call the culture complex: that is, the public ordering of the relations between particular kinds of knowledges, texts, objects, techniques, technologies and humans arising from the deployment of the modern cultural disciplines (literature, aesthetics, art history, folk studies, drama, heritage studies, cultural and media studies) in a connected set of the apparatuses (museums, libraries, cinema, broadcasting, heritage sites, etc.). The historical and geographical distinctiveness of this complex consists in its organisation of specific forms of action whose exercise and development has been connected to those ways of intervening in the conduct of conduct that Foucault calls governmental. Foucault, it will be recalled, characterises governmental power as the result of a process that, in the West, 'has led to the development of a series of specific governmental apparatuses (*appareils*) on the one hand, [and, on the other] to the development of a series of knowledges (savoirs)' (Foucault 2007: 108). The value of characterising a specific ensemble of knowledges and apparatuses as parts of a culture complex depends on being able to show that this complex brings together persons, things and techniques – ways of doing and making – in distinctive compositional configurations that give rise to, exercise and perform historically distinctive forms of action by producing distinctive techniques of intervention into the conduct of conduct.

There is no question here of an account of culture as a transhistorical constant, a component in the makeup of all societies. It makes no sense to say that some societies lack culture where this is understood as the realm of the symbolic. And to suggest that some societies lack culture, understood, in its restrictive definition, as a set of higher intellectual and cultural forms, has been a means of exercising colonial power. There are, however, many societies where the distinctive ways of assembling specific knowledges, materials, technologies and practices associated with the culture complex, and bringing these to bear on the conduct of conduct, have not been present. It is, for example, only in its post-Enlightenment conception that the library in England emerges as a site for practices of classification and arrangement, which detached books and writing practices from earlier religious assemblages, or from their functioning as quasi-military aspects of state power against the threat of popular insurgency, to be developed as key distributional technologies for shaping the conduct of the population as whole (Summit 2008). This is not to drive an essentialist divide between religion and culture or between sovereign and governmental forms of power. Rather, it is to suggest that libraries have operated across these distinctions as parts of gatherings of different kinds that have been active in their worlds in different ways rather than as historically variable sites for the operation of transhistorical mechanisms of culture.

The main analytical wager of this book, therefore, is that the historical particularity of the ways of 'making culture' and bringing it to bear on the task of 'changing society' is worth attending to along the lines proposed above. This wager is connected to a second. It follows from what has been said so far that if there can be no general disentangling of culture from the economy and the social as universals, then so neither of these realms can be thought of as being made up of a different kind of ontological stuff from the materials that are brought together in the culture complex. This is what I take to be the central import of that school of 'cultural economy', which, taking its cue from Michel Callon's work, insists that markets and other economic phenomena cannot be secured independently of the operations of a whole series of 'market devices' in which specific forms of expertise, signs, techniques and prostheses are brought together as *agencements* (Callon 2005). While similar to assemblages the concept of *agencement* places a greater stress on the capacity for such hybrid gatherings to transform the agential capacities of the elements they bring together (McFall 2009: 51). The implications of recent explorations of the 'ontological politics' of the social sciences point in a similar direction by stressing the active role played by specific forms of knowledge, and the networks of human and non-human relations and techniques of intervention they give rise to, in shaping the organisation of the social (Law and Urry 2004; Osborne and Rose 2008). Taken together, these considerations rule out those analytical procedures that seek to decipher the interactions between culture/economy/the social in the form of relations between realms or levels whose ontological constitution is different. In their place they substitute a focus on the single-planar analysis of the interweavings between economic, social and cultural assemblages that differ from one another not in the nature of the elements of which they are comprised but in their compositional configuration. The world, as Richie Nimmo puts it, 'is populated not by classes of things ontologically distinct from each other in *a priori* ways, but by radically equivalent things separated only by particular forms of organisation' (Nimmo 2010: xii).

These, then, are the issues that I shall be centrally concerned with in Chapters 2 and 3 where I elaborate the concept of the culture complex more fully and outline the place that the analysis of culture should occupy alongside other concerns within the broader field of governmentality theory. I also discuss here the revisions to governmentality theory that have been prompted by assemblage and actor–network theory and the more general reorientations of our understanding of the relations between culture, economy and the social proposed by the 'material turn' that is currently taking place across the social sciences (Bennett and Joyce 2010). More particularly, though, I shall explore Foucault's concept of liberal government for the light it throws on the relations between culture, government and freedom. We are now, courtesy of both Foucault and Nikolas Rose (1999), accustomed to thinking of freedom as an essential part of the mechanisms of government rather than as government's outside or opposite. It is, indeed, another of those universals that Foucault puts through the mill of his historical method: freedom is not, he says, 'a universal which is particularised in time and geography' but always and only 'an actual relation between governors and governed' (Foucault 2008: 63). In the case of liberal regimes of government, he continues, 'freedom of behaviour is entailed, called for, needed, and serves as a regulator, but it also has to be produced and organised' (65). But this is also, he says, a process that involves setting limits to freedom, distributing it unequally through the social body and specifying the conditions under which alternative mechanisms of governing must be found.

However, neither Foucault nor Rose – who shares Foucault's hesitancy regarding the concept of culture[11] – pays systematic attention to the relations between culture and freedom in liberal regimes of government.[12] While there have been many attempts to fill this gap, such approaches have been, in the main, partial, concerned with particular aspects of such relations rather than proposing a framework for addressing them more generally. A third analytical wager of this book, therefore, is that approaching the culture complex along the lines I have proposed will make it possible to identify the roles played by different forms of cultural expertise, operating in the context of different cultural assemblages, in producing different kinds of freedom and distributing these differentially across divisions within the population. I shall, in addressing these concerns, interrogate the ways in which these processes of 'organising freedom' are tangled up in those of 'making culture' and 'changing society'.

I offer an overview of these questions in Part 1, and focus thereafter mainly on two forms of cultural knowledge that, in their different institutional and material entanglements, have played a crucial role in defining the coordinates across which freedom, as a mechanism of government, has been distributed across different populations: aesthetics and anthropology. These have constituted the twin cultural 'authorities of freedom' whose complex and mutable interactions have played a key role in delimiting the boundaries across which freedom is to be distributed, determining where it is a mechanism to be applied in programmes of governance and where it is to be withheld. I shall, in the first part of the book, draw on examples from both aesthetics and anthropology in laying out the general contours of my argument. I do this, first, in Chapter 2 by more fully elaborating the relations between 'making culture, organising freedom, and changing society' that are briefly sketched above. I then, in Chapter 3, illustrate these by examining some of the different kinds of 'cultural objecthood' that have been produced by the museological deployment of anthropology and aesthetics. Thereafter I engage with these two knowledges in different ways in different parts of the book.

My concerns in Part 2 focus on the operations of anthropology as a part of cultural assemblages that have distributed freedom differentially across the relations between metropolitan and colonial populations. I draw here on another distinction Foucault proposed when distinguishing between those processes of government that, seeking to change conduct via changing ideas, beliefs and opinions, work via the mechanism of the public and those that operate on conduct via the milieus that condition the interactions between living beings and their environments (Foucault 2007: 20–21). I elaborate the significance of this distinction more fully in the next chapter and then return to it in Chapters 4 and 5, where I consider the relations between anthropological fieldwork, museum collections, colonial administration and programmes of public education exemplified by early twentieth-century moments in the histories of the National Museum of Victoria in Melbourne, and of the Musée de l'Homme and the Musée des Arts et Traditions Populaires in Paris. I broach these moments from the point of view of their bearing on the relations between liberal programmes of governance and biopolitical programmes of population management.

Culture in question (ii): after critique

The second general difficulty with the concept of culture consists in the supposition that some forms of cultural practice might be distinguished from others to provide the basis for a form of critique that will function as a putative source of social transformation through its influence on the consciousness of subjects. We might take heart from the assessment that critique has now run out of steam, as Latour (2004) famously put it, or, as Luc Boltanski and Eve Chiapello (2007) have argued, that it has found its *rendezvous* in the new spirit of capitalism. Such notices of its demise have, however, proved premature so far as the relations between aesthetics and critique are concerned. These have been resurrected in Jacques Rancière's conception of metapolitics as a practice of critique that he derives from an essentially Kantian conception of the relations between aesthetics and freedom.

The central questions organising my concerns in the third part of the book hinge around the relations between aesthetics, freedom and critique considered in their relations to the culture complex and regimes of liberal government. There are two main aspects of critique that I shall be concerned with. The first consists in the attempt to produce an intellectual space that is accorded a position of transcendence – even if only historically rather than absolutely – in relation to what would otherwise be the irredeemably located, and thus inevitably particular and partisan, clash of intellectual and political perspectives. This may take the form of a *sensus communis* or, in its Anglicised versions, a 'common culture' projected beyond the schismatic effects of the division of labour; it may, in its Habermasian versions, take the form of an ideal speech community that has yet to be brought into being but which, in the meantime, provides normative protocols for regulating the public use of reason; it may take the form of Rancière's metapolitics; and it may, as I argue in Chapter 9, take the form of Bourdieu's collective intellectual whose dispensations, forever waiting just around the next historical corner, are anticipated by the sociologist.

This first set of concerns is thus focused on the production and validation of a position from which critique can be enunciated. The second set is focused around the question: how can a subject who has been shaped in the midst of relations of power move to occupy or take up such a position? There are different ways of answering this. In classical Marxism, for example, it is the direction of history itself, the momentum of the real, that moves the subject – represented by the proletariat – to such a position. Foucault envisages a different mechanism. There is a passage in *The Hermeneutics of the Subject* where he says that 'there is no first or final point of resistance to political power other than the relationship one has to oneself' (Foucault 2005: 252). It is not an abstract relationship of self to self that Foucault has in mind here nor, assuredly, one produced by the movement of history but one arising out of the relationships between different technologies of the self. Nonetheless, he says, 'the analysis of governmentality – that is to say, of power as a reversible relationship – must refer to an ethics of the subject defined by the relationship of self to self' (252). And when he searches for a model for how a subject might begin to shape itself freely rather than through a relationship of tutelage to some authority or set of moral codes, it is his largely Kantian conception of the 'aesthetics of existence' that he turns to (Foucault 1989). His contention in this regard is, I shall argue, best viewed as one amongst many instances of a distinctive form of cultural authority that guides and directs particular practices of freedom even while – and, indeed, precisely by – seeming to sever them from any form of tutelage whatsoever.

I shall, in developing this argument in Part 3, interpret aesthetics like any other form of knowledge: as a particular regime of truth caught up in the processes through which the exercise of particular kinds of power is enacted, rather than as an exception to such processes. To avoid possible confusion, my concern here is not directly with the sensate faculty of judgement or particular artistic practices but with the accounts that have been offered of such a faculty and its relations to works of art by different branches of aesthetic theory. Such accounts, I shall argue, serve to constitute a paradoxical kind of authority, an 'authority of freedom', which aims to install itself at the crossroads of the faculties in order to direct the subject's movement while, at the same time, disavowing any such interest. My concerns will be threefold. First, in Chapter 6, I trace the processes through which the aesthetic faculty acquired the capacities that enabled it to act as a lynchpin of freedom within the subject. This will involve a consideration of how this faculty was shaped through its initial associations with civic humanism and early regimes of liberal government, and the subsequent enlargement of its capacities produced by its connections to the Kantian conception of culture. It will also involve, as my second concern, a consideration of the ways in which the capacities attributed to this faculty have operated as parts of the aesthetic assemblages through which aesthetic knowledges, practices and authorities have co-mingled in various ways to induct individuals into varied kinds of self-regulating behaviour. My third concern, which I come to in Chapter 7, is to explore the paradoxical qualities of the kinds of 'guided freedom' that are constituted by the relations between aesthetics and critique exemplified by Rancière's conception of metapolitics. This will involve a consideration of the Kantian underpinnings informing Rancière's work.

Culture in question (iii): after the human/post-human divide

One aspect of my argument in Part 3 of the book will concern the influence that has been exerted, via Kant, by the legacy of Christian metaphysics on the subsequent development of Western aesthetic theory. This, in essence, derives from Kant's attribution to humanity of all of those qualities that the Christian tradition had attributed to God. This provided the basis for Kant's case, as Jane Bennett summarises it, 'not only for a qualitative gap between inorganic matter and organic life but also for a quantum leap between humans and all other organisms' (Bennett 2010: 68). The marker of the second of these dividing lines was provided by the concept of culture. Kant's conception of *'humanity as cultural agency'*, Sankar Muthu argues, contends that 'human beings are fundamentally cultural creatures, that is, they possess and exercise, simply by virtue of being human, a range of rational, emotive, and imaginative capacities that create, sustain, and transform diverse practices and institutions over time' (Muthu 2003: 7–8).

Criticisms of nature/culture dualisms are now well-established in the wake of Latour's critique of the great divide that founded the 'modern settlement' (Latour 1993), Donna Haraway's cyborg feminism (Haraway 1997) and post-humanist perspectives on the relations between animals and humans (Wolfe 2010). It is, however, Richie Nimmo who offers the most productive point of entry into these questions from the point of view of my concerns here. The attention he pays to the role played by post-Kantian theological conceptions of humanity in the development of the technical and epistemological appara-tuses that separated human and animal life as differentiated spheres of governance is particularly relevant. The category of culture, he argues, has played a central role in the production of Man as a special category, separated from and elevated above nonhumans. It has played this role, however, not as an intellectual abstraction, but through its inscription within the ontological technologies – complex assemblages of epistemic and technical productions of Man – through which the ontological separations human/nonhuman and culture/nature are enacted. The 'ontological and epistemological apparatus bound up with the concept of "culture"', he says, 'has been intrinsic to the proliferation of the knowledge practices constitutive of modernity as an historical formation' (Nimmo 2010: 15). The purification of the 'divine human domain of subject-culture' (74) by the epistemological technologies of humanism also served to produce nature as a set of processes distinct from human agency, which was to be subdued by being brought under the control of the latter through the development of a different set of epistemological and governmental technologies. It is through such reciprocal processes, he argues, 'that the very notion of human "agency" as a preponderant force acting upon a "nature" outside of itself was made sustainable, by being distinguished from "natural" forces – which were thereby denied the status of "agency" – and set on a plane of "culture" brought into being by this very distinction' (77).

The production and policing of the nature/human divide does not, however, solely concern our relations to an external nature. It also concerns our relations to the force of nature within. One of the boundary lines across which this particular work of purification has taken place is that constituted by the concept of habit. Kant on habit as the beast within:

> Habit (*assuetodo*), however, is a physical inner necessitation to proceed in the same manner that one has proceeded until now. It deprives even good actions of their moral worth because it impairs the freedom of the mind and, moreover, leads to thoughtless repetition of the very same act (monotony), and so becomes ridiculous. … The reason why the habits of another stimulate the arousal of disgust in us is that here the animal in the human being jumps out far too much, and that here one is led instinctively by the rule of habituation, exactly like another (non-human) nature, and so runs the risk of falling into one and the same class with the beast.
>
> (Kant 2006: 40)

Habit here operates as a kind of liminal zone within the human, the zone in which nature and culture contend with one another. Yet, depending on how this struggle is resolved, it also serves as a means for marking a distinction between different humanities. If culture and freedom can be attained only by combating the force of habit, some sections of humanity fulfil this task better than others. If this marks a distinction between different humanities, this is also a governmental distinction between those groups and populations that qualify as cultured in Kant's sense of possessing an aptitude for freedom and are thus able to govern themselves and those whose conduct, still unduly under the influence of habit, needs to be subjected to more disciplinary forms of regulation.

Such, in rough summary, is the place that habit has occupied in relation to the development of liberal forms of government. It has, however, proved to be a shifting liminal zone in the sense that its precise location has varied according to the position it has been accorded within different 'architectures of the person': that is, different ways of thinking the inner partitioning of persons and how different components of personhood (instinct, will, habit, memory, reflex) interact. This is true, as Jane Bennett (2010) has noted, of the respects in which Darwin's account of the role of habit in the life of worms troubles the notion of a simple dividing line between nature and human history and culture of the kind that Kant proposes. The activities of worms, Darwin argues, involve the exercise of a certain degree of freedom of choice and intelligence. His case rests principally on his observations of the manner in which worms plug up their burrows. While the plugging up itself is instinctive, the manner in which worms plug up their burrows – adapting the size and shape of the leaves, twigs or petioles they select for this purpose to the shape of the holes into which they are inserted – suggests an activity guided by intelligence in much the same way (if not to the same degree) as a man might go about the task. In this and in other ways, Darwin is led to the conclusion that the activities of worms are not the result of 'instinct or an unvarying inherited impulse' (Darwin 1881: 156),[13] or of 'a simple reflex' (156), and that they are not governed by 'inherited habit' or, indeed, by new habits 'acquired independently of intelligence' (667). Instead, and much to his surprise, his observations led him to the conclusion that worms show 'some mental power' (156) and a capacity for attention that 'indicates the presence of a mind of some kind' (242); that they show 'sexual passion' and 'a trace of social feeling' (234); and that, by dragging 'objects into their burrows first in one way and then in another, until they at last succeed' they are able to 'profit, at least in each particular instance, by experience' (690). In this way Darwin confounds the sense that an absolute distinction might be drawn between the lower and higher orders of animal life, or between animals and humans, in which, on one side of the line, behaviour is entirely governed by reflex, instinct or habit and, on the other side of that line, by consciousness, volition and intelligence.

This sense of a continuum of forms of life in which the roles of reflex, instinct, habit, consciousness, volition and intelligence are always present, both to different degrees and in different combinations, informs a good deal of the post-Darwinian literature on habit in the life sciences and in the fields of anthropology and archaeology. As such, it stands in sharp contrast to the parallel development of post-Kantian accounts of culture in which habit serves as the antithesis to the conception of culture as the process of free human self-making that is propelled forward by the example of genius and the non-repetitive forms of emulation it stimulates.

These, then, are the questions I am concerned with in Part 4 where I consider how different accounts of habit have informed the ways in which liberal technologies of government distribute the aptitude freedom unevenly across relations of gender, class and race. In doing so, I bring together the perspectives on anthropology and aesthetics developed in Parts 2 and 3 by means of two case studies. I look first, in Chapter 8, at the role played by post-Darwinian accounts of the relations between will, habit, memory and instinct considered in their relations to the late-nineteenth-century doctrine of survivals and its implications for colonial forms of government. My concerns here complement my discussion of the relationships between museums, anthropological field-work and the governance of Aborigines in early twentieth-century Australia in Chapter 4. I then, in Chapter 9, look at Bourdieu's conception of the relations between habitus and habit with a view to disclosing how these are shaped by the constitutive tensions of liberal social and political thought. My interests here focus on the ways in which he distinguishes the degrees and kinds of aptitude for freedom that can be accorded to different habitus. I show how his conception of the habitus of the members of 'archaic' societies draws on the late-nineteenth-century doctrine of survivals, and then decipher the Kantian underpinnings that inform his account of the working-class habitus in *Distinction*. I shall, for this purpose, return to some aspects of my discussion of Rancière's work in Chapter 7 where I consider the ways in which Rancière pits the authority he constructs for himself as a spokesman for the aesthetic against that of the sociologist as represented by Bourdieu. While by no means minimising these differences I am more interested, in this final chapter, in the respects in which Bourdieu's work can be read as a sociologised bid for the authority of the 'guided freedom' that the aesthetic produces as a historically distinctive form of tutelage.

****

After culture? My response is: 'no, not yet' in the sense that the concept has formed a (shifting) part of a historically specific configuration of practices that has still to be reckoned with – analytically, politically and ethically – as a set of governmental conditions and effects. These practices, though, are not what might be taken to be the 'raw materials' of culture, whether these are thought of as beliefs, customs, traditions, rituals or everyday ways of life, or as literary, musical, art or media practices. My concern is not with the properties of such practices 'as such' or 'in themselves' but with the diverse ways in which the relationships between them are orchestrated through the intersections of the knowledge and governing practices in which they are assembled.

2 Making culture, organising freedom, changing society

Let me go back for a moment to Foucault's discussion of culture in *The Hermeneutics of the Self*. Thomas Osborne, as we have seen, glosses the distinctiveness of this passage in terms of the '"veridical" twist' it proposes in contending that 'the culture of the self is also a culture of truth' (Osborne 2008: 70). Osborne's interests focus on the implications of this veridical twist for the analysis of modern aesthetic culture. Approaching this as a practice of freedom of a particular kind, he places it in a different compartment from other conjunctions of culture, truth and practices of the self. There is much to admire in Osborne's discussion, and I shall do it greater justice when I engage with it more fully in Chapter 7. I cite it here in order to highlight, by way of contrast, how I shall interpret the 'veridical twist' that Foucault brings to the concept of culture in order to bring it more into line with the principles of historical reasoning that Foucault recommends. These require that we treat aesthetic culture alongside the regimes of truth of a range of cultural disciplines, and the ways in which these work on, regulate, maintain or transform conduct through a variety of routes and mechanisms, some of which depend on practices of the self while others do not.

I shall particularly urge the need for the roles played by aesthetics in these regards to be placed alongside those of anthropology in their nineteenth- and early-twentieth-century forms.[1] As cultural knowledges that emerged and developed in tandem with one another, aesthetics and anthropology have been complexly entangled in differentiating governmental programmes working through the free government of the self versus the coercive management of the Other. They have thus constituted significant points of reference between which the articulations of the relations between freedom, truth and government associated with the modern cultural disciplines have oscillated. The exclusionary logic Foucault attributes to the concept of culture have thus to be understood as operating across the relations between knowledges rather than simply being a process that is internal to each of them (although it is this too). We can, however, take full advantage of the 'veridical twist' that a Foucauldian optic brings to the concept of culture only if we remember Foucault's insistence on the need to examine how practices of the truth are always tangled up with the material histories of specific governmental techniques and technologies.

This is not just a matter of considering the material entanglements of cultural practices within specific apparatuses (museums, libraries, etc.); it also involves a consideration of the place such apparatuses occupy in relation to broader material networks and infrastructures, including the roles these play in organising and distributing freedom.

These, in rough summary, are the issues I address in this chapter. I do so by outlining the relations between culture and the perspectives of governmentality theory in the light of the veridical, material and compositional 'twists' outlined in the previous chapter. I begin by discussing the concept of the 'culture complex', a term I have proposed to refer to a distinctive ensemble of regimes of truth, governmental rationalities, techniques of the self and other modes of intervening in the 'conduct of conduct' that has operated alongside the 'psy-complex' (Rose 1985) and other ensembles of knowledge and governing practices that have emerged over the course of the 'modern' period (Bennett 2010b, 2012). I then consider the relations between this complex and the somewhat vexed account Foucault offers of the relations between the emergence of the social and of population on the one hand and the development of governmental power on the other. I look briefly here, as matters I shall return to, at the roles of the public and the milieu as different 'transactional realities' through which culture's governmental action is routed in organising and distributing freedom across the relations between liberal forms of government and biopower and the different relations to population that these involve. I then identify the implications of these considerations for a properly historical understanding of the relations between culture and the social before turning to the ways in which assemblage theory and science studies can usefully enrich the concerns and methodologies of governmental theory. I conclude the chapter by contrasting the account it proposes of the ways in which culture's modes of action in relation to the social have been historically produced and assembled with Durkheimian conceptions of the autonomy of culture, while also outlining its implications for an ontological politics of culture.

The culture complex

I have, in earlier work, suggested that the nineteenth-century development of public museums and exhibitions of various kinds might usefully be interpreted as an 'exhibitionary complex' that paralleled the apparatuses of discipline comprising Foucault's carceral archipelago (Bennett 1988). Drawing on Antonio Gramsci's account of the ethical state, I argued that this complex constituted a 'soft' pedagogic alternative to the rigours of discipline by organising voluntary consent to bourgeois forms of hegemony. However, I also drew on two aspects Foucault's work to account for the *modus operandi* of this complex. First, I suggested that it operated through a distinctive set of knowledge/power relationships constituted by the 'exhibitionary disciplines' (geology, history, art history, archaeology, natural history and anthropology). These aimed to recruit the citizens of liberal–democratic polities as active subjects of the varied rhetorics of progress, which informed the evolutionary orderings of things and peoples produced by those disciplines. Second, taking my cue from Foucault's account of the panopticon as a specific architectural distribution of the relations between seeing and being seen that formed a part of the exercise of disciplinary power, I suggested that the exhibitionary complex effected a different organisation of such relations in producing spaces in which an assembled citizenry could watch over itself and thus, in monitoring its own behaviour, become self-regulating.[2]

While it still has its uses, the concept of an 'exhibitionary complex' suffers from a number of limitations from the point of view of my concerns here.[3] The description of anthropology, geology, natural history, art history, etc., as 'exhibitionary disciplines' arbitrarily privileges one set of institutional contexts for their deployment over others, particularly administrative ones. It also draws too sharp a dividing line between those disciplines and other cultural disciplines (literary studies, musicology) whose practices do not necessarily have a significant exhibition component and whose institutional articulations are different (the library, the concert hall). A further limitation of the concept consists in its tendency to restrict the analysis of the 'exhibitionary disciplines' to their influence on the conceptual and perceptual frames or comportments of museum visitors. John MacKenzie (2009), with this in mind, has argued that the concept throws little light on the dynamics of museums in 'settler' colonial contexts where, since they were not welcomed as visitors, indigenous peoples were not subjected to the forms of civic regulation and surveillance of the 'exhibitionary complex'. The early history of the relations between museums and indigenous peoples is, in truth, more varied than MacKenzie allows.[4] Nonetheless, his objection highlights the limiting supposition that it is only as visitors that indigenous peoples might have been affected by colonial museums. This neglects how the forms of ethnographic knowledge produced by the classification and ordering of indigenous collections within museums have been carried back to and acted on indigenous populations via their application through the networks and apparatuses of colonial administration.[5]

I propose the term 'culture complex', then, to encompass the roles played by a broader range of knowledge practices and institutions in the governance of conduct. But what kind of analytical focus does the reference to culture bring to these questions? Osborne, noting that Clifford Geertz interprets culture 'not as complexes of concrete behaviour patterns – customs, usages, traditions, habit clusters – but as a set of control mechanisms – plans, recipes, rules, instructions (what computer engineers call "programmes") – for the governing of behaviour' (Geertz, cited in Osborne 2008: 16), argues that this conceptualisation breaks down in complex, functionally stratified societies in which the economy, politics, law and culture are differentiated from one another. This means, he says, that it becomes difficult to think of culture as 'a generally co-ordinating form of symbolic activity' as it comes to be understood more 'as a particular "sphere" of social existence in its own right' (Osborne

2008: 17). It is, again, aesthetic culture that serves as Osborne's model here for an account of the autonomisation of culture that relies implicitly on Bourdieu's account of the differentiation of the cultural field from the economic, social and political fields. I adopt a different approach to the historical assemblage of culture, interpreting this not as a process driven by the struggles of artists and intellectuals for independence from the fields of economic and political power. My interest is rather with the role played by epistemological authorities of various kinds in producing new collectivities of actors and endowing these with specific capacities for acting on and changing conduct. This entails that attention should focus on the interactions between the practices that are distributed across the division that Geertz proposes between customs, usages, traditions and habit clusters on the one hand, and plans, recipes, rules and instructions on the other, as a consequence of their being brought together within a historically specific set of apparatuses. Culture, thus understood, operates in and through the conjunctions that it effects between the problem spaces that are produced by those knowledges concerned with 'customs, rituals, traditions and habit clusters' (anthropology, folklore studies, sociology) and the ways in which other cultural disciplines (aesthetics, art history) act on those problem spaces through the points of access to and means for intervening in the conduct of conduct that they provide.

One of the virtues of the concept of assemblage is the degree of plasticity it brings to our understanding of the ways in which the relations between different forms of knowledge and conduct are arranged in being contingently brought together into provisional combinations and modes of interaction in specific settings. However, it is what Deleuze has to say about the functioning of the diagram within such processes of assembling and disassembling that I want to highlight here. The diagram or 'abstract machine', Deleuze argues, acts not like a transcendental idea or an ideological superstructure but as 'a non-unifying immanent cause that is co-extensive with the whole social field: the abstract machine is like the cause of the concrete assemblages that execute its relations; and these relations between forces take place "not above" but within the very tissue of the assemblages they produce' (Deleuze 1999: 37). The diagram, as the organising principle of a power/knowledge relation abstracted from any particular use, functions in a particular manner: 'It never functions in order to represent a persisting world but produces a new kind of reality, a new model of truth. ... It makes history by unmaking preceding realities and significations, constituting hundreds of points of emergence or creativity, unexpected conjunctions or improbable continuums' (Deleuze 1999: 35).

The example Deleuze has in mind concerns the functioning of the panopticon, as the diagram of discipline, in relation to the 'concrete assemblages' of the school, workshop or army in acting as an 'immanent cause' regulating the arrangements between different categories of person, the architectural principles regulating lines of sight, and so on. The resulting conception, as Kevin Hetherington has usefully summarised it, is one of a decentered conception of the social as made up of a 'messy set of interconnections and combinations of diagrams' whose actions, and reactions, comprise 'an operation of forces that constitute relations of power within society without reducing that society to a single function of power' (Hetherington 2011: 459).

To interpret the culture complex as comprising (for the moment) an interacting set of concrete assemblages entails a similar concern with the role played by diagrammatic forces in constantly undoing and producing new realities, new fields of action in which conduct is brought under the influence of a mutating set of relations between specific cultural knowledges and the assemblages in which they are embedded. The issues at stake here can be illustrated by briefly reviewing the history of *Bildung* as an 'aesthetic diagram' that came to be tangled up in a variety of cultural assemblages. Reinhart Koselleck has traced the stages through which *Bildung*, a practice of self-formation derived initially from the forms of self-inspection and correction proposed by the Earl of Shaftesbury's modelling of the aesthetic on an inner soliloquy (Shaftesbury 1999), was subsequently given a theological inflection through its connections with German Pietism. *Bildung* thus equipped artistic practices with a reformatory capacity that was subsequently connected to a range of social and political programmes: first, through its role in training the new corpus of experts, administrators, scientists, etc., who formed the nucleus of the bureaucratic state; second, through its role in the internal forms of socialization through which the bourgeoisie – in marriage, in social life, in clubs and at home – secured a specific identity for itself; and third, through its use in programmes of public education (Koselleck 2002: 172–3). However, Koselleck offers little sense of the deeply material processes that were involved in the new forms of connectivity that *Bildung* became a part of in each of these social inscriptions. The process of hitching it to programmes of public education thus involved its articulation with and within a new ensemble of institutions (public libraries, concert halls, museums, art galleries and exhibitions) that brought new publics together with new combinations of things, texts, technologies and instruments in specially contrived architectural spaces. These spaces were also themselves parts of new forms of socio-spatial ordering associated with the moral economy of the liberal city in which *Bildung* was hard-wired into the material environment (Joyce 2003).

These new cultural assemblages were produced through the (partial) deletion of earlier assemblages in which the relations between people and things, between publics, objects, texts and instruments, were differently configured. The material economy of the nineteenth-century city of culture thus depended not only on a new partitioning of urban space but on the severance of the nexus of the relations between people and things that had been inscribed in the quite different institutional nexus of the spa city and its practices (Borsay 1989). It also depended on the relocation of art objects from private settings (aristocratic and royal households), detaching these from their decorative functions or from their role in the spectacularisation of sovereign power to assemble them in new public contexts where, in being endowed with new capacities, they were refunctioned for new purposes (Rees-Leahy 2009). The deployment of the aesthetic diagram of *Bildung* in art museums thus trans- formed works of art into resources for developing a new in-depth interiority on the part of the subject. This, as I go on to show in the next chapter, opened up an inner space within the person in which a disposition of self- reform of the kind required by *Bildung* could be constructed. The relationship between new forms of design and the role of musical instruments, particularly the piano, in restructuring the bourgeois household also proved critical in reassembling the home as a space for the fashioning of the new forms of interiority associated with *Bildung* (Umbach 2009). While partly modelled on these arrangements, the translation of *Bildung* into civilising programmes in colonial contexts involved distinctive co-minglings of persons, clothes, texts and implements in the self-civilising work that Australian Aborigines, for example, were enjoined to perform under white tutelage in specially constructed environments (Lydon 2005).

Here, then, are some pointers to how the mutating connections between aesthetics and a range of new cultural institutions spilled out beyond those institutions to provide historically distinctive resources and strategies for intervening in the 'conduct of conduct' that were just as consequential in colonial Australia as in the middle-class suburbs of Berlin. This is not to suggest that the aesthetic diagram constitutes a principle governing the whole cultural field after the fashion of Deleuze's construal of the relations between panopticism and discipline. To the contrary, the institutions comprising the culture complex have constituted the locales for different diagrams that, derived from different knowledge practices, have configured the relations between persons, texts, instruments, objects and spatial arrangements differ- ently, laying out the architecture of the person in varying ways by opening it up to different practices of the self guided by different epistemological and ethical authorities.

Let me give a couple of examples derived from recent interpretations of post-Kantian versions of the aesthetic diagram considered in its relations to, in the one case, the discourses and institutions of modernity and, in the second, to those of tolerance. The key to understanding what William Ray calls the 'logic of culture', interpreted as a modern mechanism of person formation, consists in the tension it establishes between 'culture' understood as, on the one hand, 'the shared traditions, values, and relationships, the *unconscious* cognitive and social reflexes which members of a community share and collectively embody', and its post-Kantian usage to refer to the 'the *self-conscious* intellectual and artistic efforts of individuals to express, enrich and distinguish themselves, as well as the works such efforts produce and the institutions that foster them' (Ray 2001: 3). Culture, he argues, inscribes our identities in the tension it produces between inherited and shared customs and traditions on the one hand, and, on the other, the restless striving for new and distinguishing forms of individuality: 'it tells us to think of ourselves as being who we are because of what we have in common with all the other members of our society or community, but it also says we develop a distinctive particular identity by virtue of our efforts to know and fashion ourselves as individuals' (3). Culture, on this view, is a mechanism that takes issue with habit, tradition, custom, superstition: these are the 'adversary to be overcome before we can realise our full humanity' (16).[6] It thus initiates a process of critique through which the individual extricates him or herself from unthinking immersion in inherited traditions in order to initiate a process of self-development that will result in new codes of behaviour. These codes – in being freely chosen rather than externally imposed, and in meeting the requirements both of reason and of individual autonomy and expression – distinguish those who have thus culturally reformed themselves from those who remain unthinkingly under the sway of habit, custom or tradition.

This 'logic of culture' has played a significant role in Western exhibition practices from the nineteenth century through to the present. Jonathan Crary underlines the significance of the issues at stake here in noting the apprehensions that were generated, in the late nineteenth century, around the new forms of distracted and automatic forms of attention associated with industrial production and the development of new forms of popular visual entertainment. The fear was that, owing to the association of the habitual with instinctual rather than rational procedures, modes of perception that had become routinised 'no longer related to an *interiorisation* of the subject, to an intensification of a sense of selfhood' (Crary 2001: 79). They were therefore inimical to the production of those forms of tension and division within the self that are required for the machinery of culture to take a hold and be put to work within a dialectic of self-development in which individuals renovate and distinguish themselves from the common mass by disentangling their selves from the weight of unconscious inherited reflex and traditional forms of thought, perception and behaviour. It is, then, not surprising that, as an instrument of culture, the modern art museum has been committed to a programme of perpetual perceptual innovation. It has sought to disconnect vision from falling, so to speak, into 'bad habit' by critiquing not only the distraction of attention associated with popular visual entertainments but also the flagging forms of perception associated with earlier artistic movements that, while once innovative and able to provoke new forms of perceptual self-reflexiveness, have since atrophied into routine conventions.

My second example is drawn from Wendy Brown's analysis of the historical transformation in the functioning of discourses of tolerance that has been produced by their contemporary articulation to programmes of diversity management. This is partly a matter of the new institutional contexts through which such discourses circulate. 'Once limited to edicts or policies administered by church and state', Brown argues, 'tolerance now circulates through a multitude of sites in civil society – schools, museums, neighbourhood associations, secular civic groups, and religious organisations' (Brown 2006: 37). Interpreting this as a shift related to a transformation in the functioning of tolerance from 'an element in the arsenal of sovereign power to a mode of governmentality' (37), Brown argues that this mode of governmentality works through the distinctions it organises between those it constructs as the subjects of tolerance and those who are its objects. These distinctions, she argues, are produced by, and work within, the disjunctions between aesthetic conceptions of culture as applied to cosmopolitan liberal subjects deemed capable of shifting their identities, values and beliefs by moving flexibly between life-styles, and anthropological conceptions of culture in their application to a range of (usually) racialised Others who are defined as being so bound to a particular culture that they are hemmed in and limited by it.

A diagram, then, is a principle that organises the relations between the heterogeneous elements that are brought together across a range of cultural assemblages, shaping the forms of action they are capable of. There is never just one such diagram; particular assemblages may, indeed, be subject to the influence of multiple diagrams. This will become clearer as I now go on to consider what such assemblages act through and what they act on.

Publics, milieus, networks and infrastructures

Let me first summarise my argument so far. My contention is that a distinctive field of government has been shaped into being through the deployment of the modern cultural disciplines (literature, aesthetics, art history, folk studies, heritage studies, cultural sociology, cultural studies) in the apparatuses of the culture complex (museums, libraries, cinema, broadcasting, heritage sites, etc.). These comprise distinctive technologies that connect particular ways of doing and making – particular regimes of practice – to regularised ways of acting on the social, which aim to bring about calculated changes in conduct by transforming beliefs, customs, habits, perceptions, etc. These changes are related to particular rationalities of government that identify the ends towards which governmental activity is to be directed, the populations whose conduct is to be managed, the instruments through which such management is to be effected and the means for monitoring their effectiveness. This is not, how-ever, to suggest that the culture complex operates on one side of a historical dyke that separates it entirely from earlier forms of power. If Norbert Elias counterposes the post-Kantian ethos of culture to the forms of etiquette cultivated by the upper classes of Germany's earlier courtly societies, he nonetheless traces significant historical affiliations between the two in terms of the longer term dymanics of 'the civilising process' (Elias 1978). The scripts of many museums are equally clearly a mix of sovereign and governmental forms of power (Bennett 2006); many contemporary cultural disciplines and apparatuses are still marked by their relations to earlier forms of pastoral power (Hunter 1988); and opera remains significantly linked to sovereignty (Bereson 2002). However, these qualifications do not affect our capacity to distinguish the relations between the culture complex and the cultural dis-ciplines as a historically distinctive ensemble of power relations and practices any more than the continuation of sovereign power alongside discipline and governmental power invalidates Foucault's identification of these as different modalities of power that, as he frequently emphasized, were often complexly co-mingled (Foucault 1991: 101–2).

Foucault elaborates this point at the beginning of *Security, Territory, Population* where he stresses that the relations between the mechanisms of sovereignty, discipline and security are not ones of historical succession. They are rather, he says, ones in which the principles of sovereignty and discipline are transformed as they continue to exist alongside, but are subordinated to, the *modus operandi* of security as the new dominant or coordinating principle of power. The defining principle of the mechanisms of security, Foucault goes on to say, is its concern with 'population as both object and subject' (Foucault 2007: 11). This echoes an earlier formulation where Foucault argues that, in the case of governmental power, 'the population is the subject of needs, of aspirations, but it is also the object in the hands of government, aware, *vis-à-vis* the government, of what it wants, but ignorant of what is being done to it' (Foucault 1991: 100). The account of the emergence of population as the primary target of, and conduit for, the exercise and relay of governmental power on which these contentions depend is notoriously evasive on a number of key historical and theoretical points.[7] However, it is how Foucault relates the distinction between security's concern with population as object and as subject to the different ways in which it is constituted as, in the first case, a component of a milieu and, in the second, as a part of a public that concerns me here. Drawing on Georges Canguilhem's discussion of the transformations through which the concept of milieu was adapted from its original use in Newtonian physics to become a key term in the life sciences and, subsequently, in the social sciences (Canguilhem 2008), Foucault interprets milieu as a set of natural and social 'givens' through which there circulates a combined set of effects on all of those who live in it.[8] It is in this regard, he argues, that the milieu presented itself as a new conduit for the action of government:

> Finally, the milieu appears as a field of intervention in which, instead of affecting individuals as a set of legal subjects capable of voluntary actions – which would be the case of sovereignty – and instead of affecting them as a multiplicity of organisms, of bodies capable of performances, and of required performances – as in discipline – one tries to affect, precisely, a population. I mean a multiplicity of individuals who are and fundamentally and essentially only exist biologically bound to the materiality within which they live. What one tries to reach through this milieu, is precisely the conjunction of a series of events produced by these individuals, populations, and groups, and quasi natural events which occur around them.
>
> (Foucault 2007, 21)

The transformation of the legal subjects of sovereignty into elements of a milieu is made clearest, Foucault argues later, when 'men are no longer called "mankind (*le genre humaine*)" and begin to be called "the human species (*l'espèce humaine*)'" (75).[9] It is in this context that he introduces the concept of public as, alongside milieu, a second conduit through which governmental practices act on population:

> From one direction, then, population is the human species, and from another it is what will be called the public. Here again, the word is not new, but its usage is. The public, which is a crucial notion in the eighteenth century, is the population seen under the aspect of its opinions, ways of doing things, forms of behaviour, customs, fears, prejudices and require-ments; it is what one gets a hold on through education, campaigns and convictions. The population is therefore everything that extends from biological rootedness through the species up to the surface that gives one a hold provided by the public. From the species to the public; we have here a whole field of new realities in the sense that they are the pertinent elements for mechanisms of power, the pertinent space within which and regarding which one must act.
>
> (75)

Foucault clarifies what he regards as distinctive about the eighteenth-century usage of 'public' when he contrasts its functioning within the absolutist prin-ciples of *raison d'État* with its operation in the later apparatuses of police. The production of public opinion forms a part of the politics of truth of the former in the sense that sovereign power must act on the ways of thinking and doing of those it recognises as juridical subjects. Nonetheless, Foucault argues, it relates to such subjects in a purely passive mode, 'giving individuals a certain representation, and idea, ... imposing something on them, and not in the least ... actively making use of their attitudes, opinions, and ways of doing things' (278). The functioning of population here, he concludes, 'had not yet entered into the reflexive prism' (278) that was to characterise it within the eighteenth-century apparatuses of police. This 'reflexive prism' consists in the relations between education and occupation. In contrast to the regulation of the relations between statuses associated with the operation of sumptuary laws, for example, police distributes people between different occupations and trains them to perform the activities associated with those occupations. It comprises a 'set of controls, decisions and constraints brought to bear on men themselves, not insofar as they have a status or are something in the order, hierarchy, and social structure, but insofar as they do something, are able to do it, and undertake to do it throughout their lives' (321). The 'reflexive prism' to which police subjects the principle of population is one that makes active use of attitudes, opinions and ways of doing things by tailoring these to the needs of the state, and thereby, to increasing the happiness and wellbeing of its subjects.

Here, then, the legal subjects of sovereignty have become the trained agents of specific occupational competencies that are actively mobilised by state authorities to secure the ends of police: to produce an adequate population in relation to the resources and territory of the state; to secure the necessities of life; to achieve the health and wellbeing of the population; to train and distribute the activity of the population to secure these ends; and to circulate the products of that activity through the provision of appropriate material networks and infrastructures, and the rules and regulations for their use. The public, in this conception, functions as a means for acting on population as a 'technical-political object of management and government' (70). As such, it is distinct from both the juridical subject of rights associated with the public in sovereignty and from the later nineteenth-century subject of interests associated with the liberal public sphere. There is, in contrast to the latter, no space within the system of police, for practices of government that will seek to work through the free activity of individuals as an alternative to directive forms of state activity. We are still in a set of governmental coordinates in which, to borrow the terms of Timothy Blanning's economical contrast, the 'culture of power' – that is, the 'representational culture' of the *ancien régime* that made, first, aristocratic and, later, absolutist power publicly manifest – has not yet ceded any space to the 'power of culture', understood as a set of technologies of self-shaping (Blanning 2002).

The principle of police when considered in relation to the arts, Foucault thus argues, is that of splendour understood as 'the visible beauty of the order and the brilliant, radiating manifestation of a force' (Foucault 2007: 314).[10] There is no question here of their use, to recall Yudice's term, as a resource for shaping the aptitudes through which subjects might govern themselves. This is clear in the early texts on police that Foucault has alerted us to. 'The division of this kingdom into parishes', Jonas Hanway writes in his 1775 text *The Defects of Police*, 'implies the communication of a knowledge of the circumstances of individuals, as they are under the influence of the *clergy*; the power of the *lord of the manor*; the authority of the *justice of the peace*; the affluence and benignity of the *gentleman*, and the benevolence of the *philosopher*' (Hanway 1775: 9). Ascribing the responsibility for the breakdown of this apparatus and the consequent increase in vice and immorality to the corrupting influence of popular entertainments and spectacles, the only forces Hanway can initially propose to counter these tendencies are religion or the gallows (or transportation). However, a third alternative occurs to him when recommending that 'the state of the stage' might be considered 'as a part of *police*' understood as part of a 'plan or design' for the cultivation of moral conduct (50). The stage can play this role, however, only provided, first, that its dramas are themselves virtuous and, second, that it is connected to other elements of the apparatus of police. While 'the good government of the lower classes of the people, might receive a very great and essential advantage, and the stage be tributary to the improvement of their morals', this is largely because its lessons will be relayed to them by the 'middle ranks' who 'are so linked in the chain of society, whatever they learnt, would be gradually communicated to their inferiors; and descending into the hearts of the indigent, and distress'd, render them *happy*, as it made them *wise*' (48). A similar logic is evident in Johan Peter Frank's *A System of Complete Medical Police*, written between 1779 and 1788, which connects the ruler's responsibility in providing popular entertainments to the requirement of maintaining a healthy economy of the body. Like Hanway, while Frank sees the theatre as a possible adjunct to the mechanisms of police, this is so only if it delights the mind or puts the body into 'moderate motion' (Frank 1976: 170) and eschews the psychologically and physically debilitating consequences of the excessive emotional and physical tensions induced, respectively, by 'Wertherism' and by tragedies.

The logic of police, then, opens up the possibility of retooling the theatre as an instrument of statecraft, directing it towards the end of securing the health and wellbeing of the population conceived, from the perspective of *raison d'État*, as a national resource. There is, though, no sense of the theatre as part of a related set of apparatuses acting to secure the same purpose; nor is there any sense that questions of freedom are implicated in the ways such apparatuses work. It is only in relation to 'the interplay of freedom and security' that constitute the later 'economy of power peculiar to liberalism' (Foucault 2008: 65) that such questions come into view. Foucault addresses these in his course of lectures on biopolitics delivered the following year. Freedom, Foucault argues, 'is never anything other – but this is already a great deal – than an actual relation between governors and governed' (63). But it is a relation whose form is subject to historical variation between different systems of rule. In the case of *raison d'État*, the limits of government are specified by means of 'a division within subjects between one part that is subject to governmental action, and another that is definitely, once and for all, reserved for freedom' (11). Liberalism works through a different 'game of freedom' (65), one that makes up different kinds of freedom – of opinion, of movement, the freedom to buy and sell – as the mechanisms through which it operates. If this is a form of governmental reason that poses the question of the limits of government as one of 'how not to govern too much' (13), it simultaneously multiplies the zones and circuits through which government works by producing the freedoms it needs for its excercise. Rather than a given that has to be respected, 'freedom of behavior is entailed, called for, needed, and serves as a regulator, but it has also to be produced and organised' (65). There are, as a consequence, no parts of the subject, no zones of personhood, that liberal government should not be concerned with if it is to produce those conditions and capacities for freedom that are required if government is not to be excessive. Liberalism is a form of governmental reason that multiplies what I shall call 'authorities of freedom': that is, forms of expertise that specify the limits of freedom in terms of the zones of conduct and the populations to which its principles shall apply, define the variable forms in which freedom is to be produced, and equip those designated as its subjects or bearers with the capacities needed to enact and perform the freedoms that are required of them in accordance with calculations based on the interplay between freedom and security, between individual and collective interest.

The culture complex is a historical product of this governmental rationality. It comprises a connected set of institutions and knowledge practices that provide the conditions for the free production and shaping of selves, identities (individual and collective), beliefs and forms of conduct for specific populations. Its emergence is connected to that of the liberal public sphere provided that this is understood not, as Jürgen Habermas (1989) proposes, as a set of institutions within which, through reasoned debate, a set of opinions was formed and brought to bear critically on the exercise of state authority. The task is rather to reconsider the liberal public sphere as the production of a new ordering of the relations between liberal forms of government and the social after the fashion of Foucault's account of the emergence of civil society.[11] Disputing nineteenth-century philosophical accounts of civil society as 'a reality which asserts itself, struggles, and rises up, which revolts against and is outside government or the state' (Foucault 2008: 297), Foucault construes such accounts as part and parcel of the processes through which civil society was produced and enacted as a new 'transactional reality' regulating the relations between government and the economy as an aspect of a new liberal governmental technology. The production of civil society, he argues, provides an ensemble of relations within which 'economic men, must be placed so that they can be appropriately managed' (296) as parts of a technology of government that respects the autonomy of the economic laws and processes over which, nonetheless, it must preside.[12]

The liberal public sphere has similarly to be understood, in Blanning's terms, as 'both the creation and the extension of the state' (Blanning 2002: 13), which thereby produces a series of spaces outside itself and, to varying degrees, independent also of religious authorities,[13] in which individuals and groups are equipped by new cohorts of authorities with the competencies they need to exercise the freedoms they are called on to perform. The public library, Patrick Joyce thus argues, played a crucial role in performing liberalism. In making ignorance aware of itself and providing the means for overcoming this limitation it provided 'the light of publicity' – that is, the free availability and circulation of knowledge – needed for participation in liberal processes of opinion formation (Joyce 2003: 98–143). At the same time, the presence of the library at the municipal level provided a key mechanism through which liberal techniques for ruling at distance through local or community identifications operated. Similarly, in contrast to the prescriptive forms of social ordering of the court, group membership and identities are (seemingly) freely chosen through the cultural activities individuals elect to participate in. In this sense, the institutions of the culture complex perform what William Ray usefully calls a form of 'social triage' through which individuals seem to '*sort themselves into groups*' (Ray 2001: 91).

It is, however, equally important to connect the institutions of the culture complex to the development of those new material infrastructures that, in contrast to those which instituted and performed the principles of sovereignty (Mukerji 1997, 2010), proved so crucial to liberalism in creating the environments in which the capacities of freedom could be produced and exercised.[14] There is now a considerable body of historical work on the role played by the development of a range of technical infrastructures – paving, lighting, public sanitation, water – in shaping environments that, as Chris Otter puts it, provided 'the tools to be decent, healthy, sober, and self-governing', thus creating 'an apparatus within which the self could be worked on, and through, as an autonomous agent' (Otter 2008: 19).[15] Governance here has to do with structuring the material conditions in which it is possible for others to act, producing and organising freedom as a particular kind of human agency that depends on the agential capacities of material actors as much as it does on those of individuals or human collectives. Otter is particularly concerned with the relationship between particular kinds of visual agency and moral capacities engendered by the use of plate glass and public lighting in a range of nineteenth-century civic contexts, exhibitions and libraries. These 'vital materialities', in Jane Bennett's sense, 'stimulated and sustained a panoply of individual visual norms and capacities: productive attention, sensory aware-ness, urban motility, social observation, private reading', which, 'when meshed with the acting body, actualised optical capacities that made normal and durable the autonomous, rational, judging, distant practices of the liberal subject' (259). They translated the capacities for self-watching of the kind extol-led by Shaftesbury and Adam Smith – which I discuss in Chapter 6 – into practicable technologies of government with an extended, albeit not universal, social reach. The differential distribution of these new kinds of visual agency also served, Otter argues, to define some groups against others. The liberal subject's visual capacities were pitted against the desensitised and dysfunctional sensory apparatus of the poor whose lack of an adequately differentiated visual capacity disqualified them from the exercise of liberal agency, thus rendering them governable through the manipulation of their milieus.[16] This was true, too, in the sanitary realm through the networks that connected the civilising spaces of the library, museums and exhibitions to the role of the cubicle, in both private homes and public baths, in providing an opportunity and nurturing a capacity for individualised forms of self-inspection while simultaneously distributing these unevenly across classes (Crook 2007).

There are limits, however, to the generalisations that can be drawn from these nineteenth-century examples. The forms of spatial organisation associated with institutions like the art gallery, public museum and library, Clive Barnett (1999) has argued, do not apply well to those produced by broadcasting and electronic media that operate across deterritorialised networks of commu-nication. The objection is a valid one, raising important questions about the need for scales of analysis that cannot be contained within the limits of 'closed assemblages' or, indeed, those of the nation state which provided the main point of reference for nineteenth-century conceptions of the social. The transnational networks and flows that 'global cities' participate in, and pro-duce, particularly through their connections to airports, telecommunications and digital infrastructures, confirms the need for forms of analysis that can address the multi-scalar inscriptions of cultural apparatuses of the kind that Barnett proposes (McNeill 2010). However, these qualifications do not detract from the main burden of my argument concerning the need to take account of the ways in which the institutions and practices of the culture complex have always to be placed at the junctions of intersecting infrastructures, socio-material networks and public spheres. Work on the early history of cinema as a governmental apparatus has thus traced its manifold connections to restricted circuits for the production of aesthetic forms of self-regulation and inspection; to broader distributional circuits aimed at spreading liberal norms through the schooling and factory systems; and to systems, operating at the intersections of social surveys and audience studies, for inspecting the conditions of existence of problematic populations – defined in terms of both race and class – warranting more directive forms of governmental intervention (Anderson 2008; Grieveson 2009; Wasson 2005, 2008). Work on broadcast television similarly emphasises not just the, usually, more informal forms of expertise that are specific to it – such as those associated with reality TV and 'make-over' programmes – but its relations to the intricate set of infra-structures that produce and sustain the home as a key technology of liberal governance (Ouellette and Hay 2008; Hay and Andrejevic 2006). Similarly, the emergence, in colonial contexts, of vernacular cultural fields orientated to the production of new liberal forms of subjectivity is also connected, in the case of India, to the modes of colonial spatialisation produced by the intro-duction of transportation and irrigation structures and the divisions between Hindu and Muslim populations – divisions between those included and those excluded from liberal forms of subjectivity – that these engendered (Goswami 2004).[17]

It is particularly in relation to populations that are excluded from the forms of self-action identified with liberal forms of subjectivity that the logic of government via milieus comes into play. Although Foucault does not pay them any particular attention, the associations between government, popula-tion and milieu have a distinctive currency in the history of French colonial practices. They came into prominence, Alice Conklin argues, in the 1890s and remained important through to the 1930s when earlier conceptions of France's unique civilising mission gave way to those of *mise en valeur* – that is, of rational and regulated economic development. This was to be effected by action on the social conditions (medical, housing, etc.) of the colonised with a view to improving their health and efficiency as, alongside the provision of railways and other communications infrastructures, the essential means for increasing the productivity of colonial labour. This meant, as Conklin puts it, that state action was redirected from educational and civilising practices that sought to 'act on individuals themselves' towards forms of intervention intended 'to alter the social milieu in which individuals functioned' (Conklin 1997: 8). As we shall see in Chapter 5, this had significant implications for the unfolding conception, from 1928 to 1939, of the role the Musée de l'Homme was expected to play in relation to France's colonies.

We have, then, to pay attention to the different scales across which cultural assemblages operate as well as to the varied ways in which they are implicated in the tasks of managing, changing or directing the conduct of different populations. We need also to pay attention to the varied forms of agency that are involved in the processes through which the components of such assemblages are gathered together, and to the capacities and opportunities this provides for working on the social via the mechanisms of the public and the milieu.

Assembling culture, working on the social

I have drawn mainly on Foucault's later work in the foregoing sketch of the culture complex as a historically distinctive ensemble of knowledge and institutional practices. In doing so, I have stressed the multiplicity of the relations through which this ensemble enacts liberal political rationalities in making up, organising and distributing freedom differentially across the intersections of milieus, infrastructures and public spheres. These later works are, as Stephen Collier (2009) has shown, much more in harmony with the topological – or, as the term I used in Chapter 1, compositional – principles that are needed for such an analysis than are his earlier writings on discipline and governmentality. These principles, Collier argues, stress the heterogeneity of the spaces and forms in which knowledges, techniques, material forms and institutional structures are co-assembled; the heterogeneity of the networks and relations that such assemblages operate across; and the endlessly dynamic processes through which the relations between the elements within such assemblages are shaken up and reassembled into new configurations.

These aspects of Foucault's later work resonate strongly with the material and relational 'turns' that now inform a good deal of contemporary cultural analysis. Chris Gosden and Frances Larson have usefully summarised the import of these tendencies in their concept of 'the relational museum':

> Museums emerge through thousands of relationships ... through the experiences of anthropological subjects, collectors, curators, lecturers, and administrators, among others, and these experiences have always been mediated and transformed by the material world, by artefacts, letters, trains, ships, furniture, computers, display labels, and so on.
>
> (Gosden and Larson 2007: 5)

The perspective Gosden and Larson advocate has greatly enriched our knowledge of how museum collections are assembled through the interactions of varied forms of agency. Related work on 'object biographies', tracing the complex routes through which objects finally reach museums, has similarly shown how museum collections have been shaped by the agency of often quite distant actors. This has significantly revised our understanding of how indigenous peoples shaped the collections of colonial museums in deciding what they would give, and what they would withhold, from exchanges across the colonial frontier (Jones 2007). The more general significance of these intellectual orientations, however, is that of presenting the museum – and, by extension, any other cultural institution – as a point of intersection between a range of dispersed networks and relations that both flow into and shape its practices and connect it to a range of apparatuses, infrastructures and milieus.

The advantages of assemblage theory, from this perspective, consist in the pliability it brings to the analysis of such networks, flows and relations.[18] This is partly a matter of the contingent nature of the connections between the elements that are brought together in an assemblage. When Deleuze asks 'What is an assemblage?' he answers that it is 'a multiplicity which is made up of heterogeneous terms and which establishes liaisons, relations between them', stressing that its 'only unity is that of a co-functioning … It is never filiations which are important, but alliances, alloys … ' (Deleuze and Parnet 2002: 69). Manuel DeLanda, in glossing this passage, stresses the radical mobility of the relations between the elements that are brought into such alliances: 'a component part of an assemblage may be detached from it and plugged into a different assemblage in which its interactions are different' (DeLanda 2006: 10). The constituent elements of assemblages are bound together not through a lineage of shared descent, or through any intrinsic connection to the other elements with which they are co-assembled, but solely through the contingent mechanisms of connection that characterise particular moments in what are constantly unfolding processes of assembling, disassembling and reassembling. These processes, as Tanya Li argues, finesse our understanding of questions of agency 'by recognizing the situated subjects who do the work of pulling together disparate elements without attributing to them a master-mind or a totalizing plan' (Li 2007: 265).

The second attribute of assemblage theory I want to comment on I take from Deleuze and Guattari's characterisation of an assemblage as, on the one hand, 'a *machinic* assemblage' of bodies and things, and, on the other, 'a *collective assemblage of enunciation*, of acts and statements' (Deleuze and Guattari 1988: 88). This is not, however, a distinction between two different levels or orders, between the order of words and the order of things, or the dualities that such a distinction might subtend:

> An assemblage of enunciation does not speak 'of' things; it speaks *on the same level* as states of things and states of content … the independence of the two lines is distributive, such that a segment of one always forms a relay with a segment of the other, slips into, introduces itself into the other. We constantly pass from order-words to the 'silent order' of things, as Foucault puts it, and vice versa.
>
> (Deleuze and Guattari 1987: 87)

These formulations challenge dualistic ontologies that divide the world into separate realms (the real/its representations; the social/culture) and then seek to probe the nature and mechanisms of their interconnections. In lieu of this they propose a flat ontology in which the real is understood as comprised by the interactions between varied assemblages whose operations and inter-actions generate significant transformative capacities through the combinatorial productivity of the heterogeneous elements (things, persons, technologies, texts, etc.) they bring together. This does not, though, invalidate distinctions between culture, the economy, and the social provided, first, that these are understood as the result of historical processes of assembly and differentia-tion, and second, that the distinctions between them are not ontologised but are rather understood as publicly ordered differences between different spheres of action.[19]

Latour is helpful here in spite of appearances to the contrary given his opposition, since *We Have Never Been Modern* (Latour 1993), to the model of the two-house collective dividing the assembly of things (nature) from the assembly of humans (society) that he attributes to the settlement of these rela-tions produced by early modern science and political thought. For the concern to distinguish culture from the social as a subdivision within the assembly of humans, is, as we saw in the previous chapter, a further aspect of the 'modern settlement' that Latour has worked so assiduously to unsettle. He makes this clear in *Politics of Nature* where he recommends that we dispense with the idea of culture alongside the ideas of nature and society to focus on the pro-cesses through which humans and non-humans are assembled into collectives whose constitution is always simultaneously natural, social, cultural and technical. Yet Latour qualifies this position when he goes on to argue that while the division between nature and society as incommensurable realms has no valid epistemological foundations, it has had and continues to have real historical force if understood as referring not to 'domains of reality' but to 'a quite specific form of public organisation' (Latour 2004a: 53).

Similarly, in *Reassembling the Social* (2005), Latour is less iconoclastic in relation to the concept of the social than in some of his earlier formulations (Latour 2002, 2004). The central difficulty, he argues, lies not in the con-cept of the social if this is thought of as a stabilised bundle of connections between human and non-human actants that might be mobilised to account for some other phenomenon: the connections between the middle classes, works of art and relations of class distinction for example.[20] Rather, problems arise when the social is also thought of as a specific kind of material – as if there were, as Latour puts it, a distinctive kind of 'social stuff' that can be dis-tinguished from other 'non-social' phenomena and then be invoked, as a social structure, to provide an explanatory ground in relation to the latter (Latour 2005: 1–4). In place of this conception of the social as a set of realities that function as the 'drivers' of other phenomena (as in the forms of the sociology of science that science studies pitted itself against), Latour recommends that the social should be thought of as an assemblage of diverse components brought together via a work of connection on the part of a varied set of agents. There is a readily perceptible case for seeing culture as made up not of a distinctive kind of 'cultural stuff' (representations) but as a provisional assembly of all kinds of 'bits and pieces', which are fashioned into durable networks whose interactions produce culture as a specific kind of public organisation of people and things – a culture complex.

Yet to grant this is simultaneously to call into question the preoccupation of a good deal of both cultural sociology and cultural studies with how to fathom the connections between two different realms accorded different ontological statuses. Such problems simply fall away as, in their place, there emerges a set of questions focused on the ways in which different kinds of cultural actors operate in and across different kinds of publicly instituted socio-material assemblages. Miles Ogborn's account of the role played by varied forms of 'script and print' in the economic and diplomatic practices of the East India Company is a good illustration of such concerns. Examining the varied forms of writing – diplomatic, administrative, economic – that circulated between Britain and the East India Company in the sixteenth and seventeenth centuries, Ogborn interprets these as 'material mobile objects' (Ogborn 2007: 32) with a view to probing 'their different modes and patterns of movement in terms of the social relations that are constituted around and through them as they are made, made mobile, transferred, and make the world' (42–3). His concern is with the materiality of the scripts he deals with, treating these as inscriptions whose effectivity is produced via the networks through which they travel, and the centres of calculation and of action (economic or diplomatic as the case may be) in which they are combined with other inscriptions and instruments of action.[21] This also means attending to the ways in which, after the fashion of Michel Callon's account of the role of various *agencements* in equipping economic actors with specific capacities (Callon 2005; also McFall 2009), persons had to be equipped with the capacities required of them if diplomatic letters were to have anything remotely resembling their intended effects. Such matters cannot be deduced from the analysis of an abstracted set of textual properties. Rather, they require that close attention be paid to the mechanisms that were put in place around such letters in order to 'produce a strong and effective relationship at a distance' (Ogborn 2007: 38).

It also entails attending to the role played by specific authorities in equipping specific textual corpora and their readers with new potentialities and capacities. Mary Poovey has examined Romantic and post-Romantic criticism (Wordsworth, De Quincy, Ruskin) in these terms, focusing on their role in disrupting the continuum that had earlier characterised the relations between informational and fictional writing. She sees the exercise of these new forms of critical expertise as paralleling the processes through which the authority of equally new forms of economic expertise was produced by disconnecting claims to factual adequacy in economic reporting and analysis from novelistic fiction or poetic fantasy. This was reciprocated by the production of literary writing as part of a new binary opposition (the literary versus all other kinds of script and print), which distinguished it from and, ideally, elevated it above both all forms of informational writing and inferior imaginative forms.

Literary writings were thus distinguished, *inter alia*, by their tendency to 'call attention to their own artistry as one means of distinguishing between the kind of value they mediated and its antithetical counterpart, the monetary value associated with commodification and the market' (Poovey 2008: 30). The production of literary writing was not, though, Poovey insists, merely a formal matter; it was a process that depended on equipping both texts and readers with new capacities by enrolling both within new institutional ensembles.

Poovey's account of the differentiation of these two forms of writing and their institutional inscriptions is a telling example of the more general processes involved in the historical differentiation of culture as a distinctive ensemble of knowledge and institutional practices considered in its relations to the economy and the social. The aspect of her account I want to emphasise consists in the attention she pays to the simultaneity of these processes as reciprocal differentiations arising from the parallel activities of the newly emerging nineteenth-century forms of both economic and literary authority. This contrasts tellingly with the periodisation Timothy Mitchell proposes in construing the economy as the product of 'certain twentieth-century practices of calculation, description, and enumeration in new forms of intellectual, calculating, regulatory, and governmental practice' (Mitchell 2002: 118). However, it is not solely with regard to questions of periodisation that the two accounts differ. For while Mitchell sees the 1930s to 1950s as the period when 'the economy' as 'the realm of a social science, statistical enumeraton, and government policy' is finally produced, he also interprets this as a distinctive moment in a longer historical process through which the economy, in being produced as a separate realm, also operated as a key point of reference against which other spheres were differentiated and in relation to which a new division between the real and its representations emerged. As statistical representations of 'the economy' began to circulate publicly, then so it became possible to 'conceive of the gap that seems to set this circulating body of information off from the processes and activities it refers to' as a 'divide between two worlds, a sphere of figures, numbers, facts, and trends on the one side, and the world to which these refer on the other' in which the latter serves as 'the realm of the material, the real' (103).

It was only when the economy had been fixed into place in this way that it became possible for a number of other spaces to be conceived 'in terms of their relations to this hermetic field: the sphere of politics or the state; the sphere of law (previously at the centre of political economy); the sphere of science and technology; and the sphere of culture' (82). It is this suggestion that the production of culture only followed on once 'the economy' had thus been fixed into place as a sphere from which that of culture could be differentiated that I question. To the contrary, the task is to examine the parallel historical development of economic, social and cultural knowledges, the forms of expertise these have engendered, the intrication of the relations between the spheres of action they have produced, and the capacities they have bestowed on human and non-human actors.

Let me go back to my earlier account of the diagram of *Bildung*. The processes of producing the aesthetic that I discussed provided a new set of resources, means and mechanisms for acting on conduct so as to regulate or modify it. These constituted ways of making culture and organising freedom that were simultaneously ways of acting on society so as to change it. This was not, however, a matter of acting on society as a given. Account has rather to be taken of the different 'working surfaces on the social' through which the varied powers that were attributed to aesthetic culture were exercised. Such conduits for culture's action consisted in the ways that the relations between specific social groups – typically classes – were defined as in need of remedial action, and in the architectures of personhood that identified specific vectors of conduct (the relations between habit and the will, for example) as the targets for culture's reformatory intervention. These processes of making culture had their own specific histories, arising from the exercise of distinctive forms of expertise. Rather than depending on some prior fixing of an economy/real, culture/representation division, they were implicated in the processes through which such separations were produced.[22] They were simultaneously in competition with alternative, equally specific processes of making culture and changing society, which drew on different institutional ensembles and knowledge formations and operated through different working surfaces on the social.

The late-nineteenth-century development, in Anglophone contexts, of pre-historic archaeological, natural history and ethnological collections thus provided alternative assemblages of texts, objects, animals and persons. These laid out the person as a series of developmental layers to provide templates for ways of fashioning the self that were linked to evolutionary conceptions of personhood. These constituted a different set of working surfaces on the social, constructed in accordance with the racial coordinates of evolutionary time, which differed from those provided by the reformatory practices of the self derived from the principles of *Bildung* (Bennett 2004). Yet if, in the British context, aesthetic and anthropological productions of culture thus pulled in different directions, their relations within late-nineteenth- and early-twentieth-century German ethnology museums were more complementary. In assembling artefacts from non-European cultures in non-evolutionary frameworks as a testimony to the manifold differences of humanity, such museums were intended to serve as adjuncts rather than as alternatives to *Bildung*. The comparative analysis of mankind's many variations they made possible was meant to prompt the German middle classes to probe their own cultural particularity and thence to fine-tune their self-development in the light of their awareness of their distinctive position *vis-à-vis* other peoples (Penny 2002). The conception of *cultures* as plural that emerged from within this understanding of *Bildung* played a key role in reformatting the social as a set of non-hierarchical, equivalent differences. This key historical shift was later carried into the English-speaking world by Franz Boas in the course of translating the lessons of his training in German ethnology museums, and his deep formation in the culture of *Bildung* (Bunzl 1996), into what was to prove a significant, albeit incomplete, departure from the evolutionary and racial hierarchies that informed the politics of difference in early-twentieth-century America.[23]

The working surfaces on the social through which culture operates are thus produced by the ways in which the social is formatted by particular knowledges and techniques so as to lay it out for particular kinds of action and intervention.[24] In the cases of contemporary policies concerned with 'governing differences' within the diagram of tolerance, these surfaces are constructed through the ways in which different ethnicities are located in relation to one another within the context of particular discursive orderings of difference, the construction of the relations between these and particular civic goals and the instruments for action and intervention that these suggest. Multiculturalism constructs one such 'working surface on the social'. It does so by organising the relations between the state and ethnically diverse populations as ones in which the latter are treated as being made up of individuals who belong to different groups (Hage 2008). Its distinguishing feature, however, is that the state relates to and deals with those individuals as individuals with particular sub-national cultural identities rather than recognising or entering into contract with those groups as such.

This contrasts with the relations that states in 'settler societies' enter into with indigenous populations where it is the community as such that constitutes the party to the contracts the state organises for managing differences. This is the case, for example, with the Shared Responsibility Agreements that the Howard government in Australia developed as a part of its Northern Territory Intervention (NTI). These agreements, as Rebecca Lawrence and Chris Gibson summarise them, 'oblige Aboriginal communities to conform to a series of specified disciplinary practices (such as improving personal hygiene, maintaining clean households, preventing school truancy and achieving literacy targets) in order to receive access to health care and other basic social services and supplies' (Lawrence and Gibson 2007). As such, they effect a more formalised insertion of the community within the distribution of governmental rewards and sanctions than do multicultural policies. Nikolas Rose (1996) rightly notes that the latter construct a place for communities as significant points of relay in the relations between the state and individual citizens. They do not, however, typically substitute the community for the individual in contractual arrangements which displace the state/individual nexus that constitutes the key couplet of liberal–democratic polities. This distinctive governmental logic, in which action on Aboriginal milieus partly complements and partly displaces educational and cultural campaigns, entered a new phase in 2011 when the Gillard Labor Government endorsed the principles underlying the NTI by withdrawing state benefits from families whose children do not regularly attend school. This targeted the community/family nexus as a key point of intervention into the management of Aboriginal conduct.

The extension of such forms of intervention into Anglo-Celtic or multicultural Australian families/communities or, indeed, urban Aboriginal populations (they only apply to remote Aboriginal communities) would be a quite different political proposition. They have, on these grounds amongst others, been strongly opposed by sections of the Aboriginal communities affected by such control orders as well as by Aboriginal public intellectuals as unwarranted and discriminatory interference with the freedom of Aboriginal Australians. Yet they have also been strongly urged and supported by other Aboriginal leaders and intellectuals – most notably by Marcia Langton, Noel Pearson and Bess Price – as a necessary form of 'tough love' needed to break the cycle of dependency that condemns Aboriginal Australians to poverty. This conservative liberal position, as Noel Pearson defines it, stakes out its grounds on the classical terrain of liberal political thought, seeing the acknowledgement of freedoms and rights going hand in hand with the distribution of responsibilities and, where the latter are lacking, justifying the withdrawal of freedoms as a necessary corrective. The logic of Aboriginal exception that is invoked here is defined in cultural rather than racial terms, and is supported as the only appropriate means of addressing the distinctive historical and cultural position of indigenous Australians.[25] If this constitutes a reversal of the logic of colonial liberalism in its differential distribution of citizenship rights across racial categories at the time of Australian Federation, it is, nonetheless, a reversal that works within the same coordinates. As such, it provides a telling example of the interlacing of public spheres, infrastructures and milieus that comprise the processes of 'making culture, organising freedom, and changing society' that have accompanied the development of modern forms of liberal government.

The ontological politics of culture

To summarise: the culture complex comprises a range of sites in which distinctive forms of expertise are deployed in 'making culture' as a set of resources for acting on society through the different routes and mechanisms constituted by its relations to public spheres, milieus and infrastructures. These processes of 'making culture' are simultaneously ones of 'organising freedom' and distributing it differentially across the population. And they are, I have argued, processes that depend on and work through the capacities they produce for the cultural actors that are their outcome. But what kinds of processes are involved here? I shall, in concluding this chapter, briefly outline how, in answering this question, I draw on the methods and procedures that have been developed within the field of science studies. I look to this quarter in view of the attention that science studies has paid to the settings – most notably laboratories – in which scientific work is conducted, and to the transformations (purifications, reductions, translations, etc.) to which scientific practices subject the materials they work with so as to produce new entities in the field of knowledge. This has significantly illuminated the distinctive powers and capacities that are folded into such entities and the ways in which such powers and capacities are then mobilised as political forces. To apply similar principles to the analysis of culture, Latour suggests, means paying close attention to the ways in which culture is made in particular settings so as to produce new entities in the field of knowledge:

> Culture does not act surreptitiously behind the actor's back. This most sublime production is manufactured at specific places and institutions, be it the messy offices of the top floor of Marshal Sahlins's house on the Chicago campus or the thick Area Files kept in the Pitts River [sic] museum in Oxford.
>
> (Latour 2005: 175)

The examples can easily be multiplied. The anthropological fabrication of culture in Sahlins's office or at the Pitt Rivers Museum is paralleled by the production of new entities in an array of custom-built settings: the artist's studio, the archives of Mass Observation, the Office of Surrealism, the production studios for reality TV, newsrooms, the art museum, film and sound archives, libraries. These settings provide the context within which new entities are produced with distinctive powers and capacities that are mobilised, in diverse ways, across different networks, by authorities of various kinds.[26] There are, of course, no essential properties connecting the production of these different entities in the field of cultural knowledge anymore than there are in the field of science. This work is rather the product of the connections that are established between the different entities that are produced in this way by secondary authorities working in equally custom-built settings (the MLA; Claude Levi-Strauss' offices at the École Practique des Hautes Études; the Warburg Institute; Raymond Williams' rooms in Jesus College, Cambridge; the Centre for Contemporary Cultural Studies at the University of Birmingham).

We can, then, see here the rudiments of an alternative to existing accounts of the processes through which the autonomy of culture is produced. Niklas Luhmann's (2000) account of art as an autopoietic system and its role, as a form of second-order observation, in producing self-monitoring and self-regulating forms of individuality is a case in point. So is Pierre Bourdieu's contrasting sociological account of the historical and social processes that lead to the autonomy of the cultural field in relation to the economic and political fields.[27] I focus here, however, on the interpretation of science studies that Jeffrey Alexander proposes in developing the version of culture's autonomy that informs his proposals for what he calls 'the strong programme in cultural sociology' (Alexander 2003: 13). As Alexander's purpose is to buttress a Durkheimian account of the role of the symbolic in constructing social relations, his account provides a useful counterpoint to the alignments of science studies, assemblage and governmentality theory that I have proposed in this chapter.

The hallmark of science studies, Alexander argues, is defined by its critique of the sociology of science. Taking issue with the sociological construction of science as being determined by underlying social structures, science studies insists on the active role of scientific practices as independent forces in their own right. By analogy, Alexander argues, the 'strong programme' in cultural sociology rebuts the determinist premises of earlier approaches in the sociology of culture to focus on the respects in which culture shapes social life rather then being shaped by it. The way Alexander then formulates this 'strong programme', however, depends on an essentially Durkheimian interpretation of science studies as an enterprise in which 'science is understood as a collective representation, a language-game that reflects a prior pattern of sense-making activity' (13). Interpreting culture analogously as an autonomous form of sense-making activity, he views its role in shaping social actions and institutions as 'providing inputs every bit as vital as more material or instrumental forces' (12). The methodological imperative Alexander derives from this account of culture's autonomy is that cultural sociology should undertake a Geertzian 'thick description' of such practices of sense making. Rather than relying on abstract formulations of culture's causality in relation to social life, it should seek to 'anchor [its] causality in proximate actors and agencies, specifying in detail just how culture interferes with and directs what really happens' (14).

Culture, for Alexander, thus comprises practices of sense making, understood largely on a semiotic/linguistic model.[28] As such, its autonomy is assumed as pre-given and historically invariant: what matters is to describe its action, not to account for the historical production of the possibility for such action. This stands in sharp contrast to science studies, which has stressed the historically specific, highly circumstantial and situation-bound mechanisms through which new objects, entities and actants in the field of science have been produced, giving rise to new forms of action on the social, arising from the alliances between human and non-human actors they make possible. Alexander's differentiation of culture from 'material and instrumental forces' also stands in sharp contradistinction to science studies, the broader field of STMS and the subsequent development of actor–network and assemblage theory in which, of course, the fabrication of new entities is precisely the result of the application of material and instrumental forces in specific settings and occasions. Chandra Mukerji provides a better distillation of the message of these traditions when she draws on their perspectives to stress the importance of 'unearthing fundamental cultural dimensions of material relations whose consequences are not so much mediated through thought or language as located in an ordering of the material world itself' (Mukerji 1997: 35). It is for this reason that I have stressed the material processes through which the historical differentiation of culture from the social, and its modes of acting on it, have been brought about by the assembly of distinctive networks of people and things. This has made possible the accumulation of distinctive powers and capacities that can be brought to bear on the conduct or persons or particular communities. This stands in sharp contrast to Alexander's account of culture's mode of action on the social as essentially concerning the influence of systems of meaning on the subjective frameworks organising forms of social interaction.

What this misses is any sense of the much stronger forms of the production of new realities that arise from the intellectual and technical procedures of particular systems of knowledge. John Law and John Urry have usefully distilled the import of this perspective when applied to the social sciences in stressing their performative role in 'enacting realities' in the sense of helping to bring into being, and to sustain, the realities they 'discover'. The role of social science research methods in producing 'public opinion' as a new agent and mode of action on the social is a case in point.[29] The consequence of this, Law and Urry suggest, is the production of multiple realities, multiple socials, arising from different traditions of the social sciences so that what has to be assessed, in considering the relations between these, are the forms of 'ontological politics' to which they give rise: that is, the particular ways of constructing and acting on the social that their theoretical and methodological procedures produce and enact through the alliances between actors of varied types that they enable (Law and Urry 2004: 404). A similar perspective informs Thomas Osborne and Nikolas Rose's (2008) account of the transition from eugenic framings of the social to ones governed by the problem space of mobility/equality associated with the variable-centred focus of the statistical trends in post-war British sociology.

It is with a similar eye to their 'ontological politics' that the operations of cultural knowledges within the culture complex need to be considered: as knowledge practices that reconfigure the relations between varied kinds of agents (human and non-human) as parts of world-making practices. This is not, though, a 'free-floating' ontological politics in which cultural knowledges are able, so to speak, to make the world as they will. It is rather a politics in which their world-making role is tangled up with, and exercised through, their relations to other knowledge practices. Their relation to the forms of the social that the social sciences produce and enact is, from the point of my concerns, particularly significant in this respect. These are among the questions I examine in the next chapter where I look at the role that museums have played as 'civic laboratories', in which the fabrication of new and distinctive cultural entities provide the means for acting on the social as variously problematised by the social disciplines.

3 Civic laboratories

Museums, cultural objecthood and the governance of the social

My primary purpose in this chapter is to explore how far methods developed in the field of science studies for the study of laboratory practices can be applied to the processes through which, in museums, new and distinctive forms of cultural objecthood are produced and mobilised in the context of programmes of civic management that aim to order and regulate social relations in particular ways.

I pay particular attention to current attempts to refashion museums so that they might function as instruments for the promotion of cultural diversity. In addressing these concerns, I also further elaborate the more general set of questions, introduced in Chapters 1 and 2, concerning the relations between specific forms of cultural expertise and processes of social management, and the historical configuration of the relations between culture and the social in those societies we call modern.

There is, of course, nothing new in the suggestion that museums are usefully viewed as machineries that are implicated in the shaping of civic capacities. To the contrary, in the late-nineteenth-century debates leading to the establishment of the Museums Association museums were commonly referred to as 'civic engines' to be enlisted in the task of managing a newly enfranchised mass male citizenry (Lewis 1989). The value of viewing them specifically as 'civic laboratories', then, depends on the light that such an analogy is able to shed on the *modus operandi* of museums as technologies that, by connecting specific forms of expertise to programmes of social management, operate in registers that are simultaneously epistemological and civic. Nor is the suggestion that there is a kinship – a family resemblance, say – between museums and laboratories a new one. It informs two recent assessments of the distinctive qualities of the modern art museum. In the first of these, Donald Preziosi characterises the nineteenth-century art museum as 'a laboratory for the education and refinement of bourgeois sentiment' (Preziosi 1996: 168) in view of its role in providing both a setting and an occasion for a new set of practices of inwardness that, in turn, were connected to the fashioning of new forms of civic virtue. In the second, Philip Fisher argues that art museums furnish a context in which what he calls portable objects – easel paintings is the case he mentions – are 'open to resocialisation and resettlement within this or that cluster of what are now taken to be similar things' (Fisher 1996: 18). It is, however, the laboratory that serves Fisher as the epistemological model for this form of portability, in view of its ability to replicate experimental arrangements of objects from one laboratory setting to another and so make possible portable, and hence generalisable, results.

That these essays should have been written by art historians is not entirely accidental. There is now a quite extensive literature in which a number of art institutions have been likened to laboratories. Although concluding that it does not fit the laboratory case as well as she had thought it might, Svetlana Alpers nonetheless finds that laboratory practice provides a useful means of probing the respects in which the artist's studio provides a means of withdrawing from the world for the purpose of better attending to it (Alpers 1998). Latour's remarks point in the same direction when he compares attempts – including his own – to free science studies from its epistemological past to the work of those who have struggled to free art history from aesthetics. Science studies, he argues, has learned a good deal from the new material histories of the visual arts that have formed a part of this severing of the aesthetic connection, especially for the light they have thrown on the multiplicity of heterogeneous elements (from the quality of the varnish and the organisation of art markets, through the history of criticism, to the organisation of the studio and the operations of art museums) that have to be brought together to make the work of art. It has also, he suggests, a good deal more to learn from the respects in which these new material histories of art have helped to displace dualistic constructions of the relations between 'the representing Mind and the represented World' (Latour 1998: 422) by demonstrating the extent to which each of the poles of such dualities is the effect of the material instruments and practices through which their relations are mediated.

While acknowledging the force that the art museum/laboratory connection has thus accumulated, I shall argue for a broader approach. This will involve, first, drawing on the perspectives of science studies and actor–network theory to look at the processes through which different types of museum are able to fabricate new entities as a result of the distinctive procedures (of abstraction, purification, transcription and mediation) through which they work on and with the gatherings of heterogeneous objects that they assemble (Latour and Woolgar 1986). It will also involve considering how the ordering of the relations between objects, and, to bring the visitor into the picture, between objects and persons, that such procedures give rise to mediate the relations between particular forms of expertise and citizens in the context of programmes of social and civic management. I shall, in pursuing these issues, distinguish the role that museums play in these regards in liberal forms of government from those associated with their role in more directive forms of rule. I want first, though, to probe more closely how far and in what respects museums are usefully likened to laboratories.

Museums as laboratories

The work of Karin Knorr-Cetina is the best route into the issues I want to explore here. This is partly because she has always been alert to the similarities between the ways in which laboratories arrange the relations between objects and persons, and similar arrangements in other scientific settings (the psychoanalytic situation) and cultural contexts (cathedrals) (Knorr-Cetina 1992). It is, however, what she says about laboratories as such – with the qualification that it is the experimental laboratory she has in mind – that I want to focus on here. The essence of laboratory practice, she argues, consists in the varied displacements to which it subjects 'natural objects'.[1] Rather than relating to these as things that have to be taken as they are or left to themselves, there are three aspects of 'natural objects' that laboratory science does not accommodate. First, she argues, laboratory science 'does not need to put up with the object *as it is*' (117) but can work with a variety of substitutes – the traces or inscriptions of objects on recording machinery, for example, or, and Pasteur's production of microbes is a case in point (Latour 1988), their purified versions. Second, since they do 'not need to accommodate the natural object *where it is*, anchored in a natural environment; laboratory sciences bring objects *home* and manipulate them on their own terms in the laboratory' (117). And third, 'laboratory science does not need to accommodate an event *when it happens*; it does not need to put up with natural cycles of occurrence but can try to make them happen frequently enough for continuous study' (117).

The conclusion she draws from this capacity of laboratories to reconfigure objects and their interrelations is as follows:

> Laboratories *recast* objects of investigation by inserting them into new temporal and territorial regimes. They play upon these objects' natural rhythms and developmental possibilities, bring them together in new numbers, renegotiate their sizes, and redefine their internal makeup. ... In short, they create new configurations of objects that they match with an appropriately altered social order.
>
> (Knorr-Cetina 1999: 27)

What Knorr-Cetina has in mind here in referring to 'an appropriately altered social order' concerns the ways in which reconfigurations of the relations between objects, and between objects and persons, that are enacted within laboratories come to be connected to, and play a role in, the reconfiguration of social relations. The example she gives concerns the effects of laboratories connected to the medical sciences and their relation to the clinic in undermining the earlier system of bedside medicine in which the authority of the physician, resting mainly on the notoriously unreliable interpretation of symptoms, was easily challenged and contested by both patients and their families. The substitution of a new set of relations between doctor and patient in which the patient was de-individualised as diagnosis came to depend on the laboratory analysis of samples altered the balance of power between them as the patient was obliged to submit to the authority of the new social collective of doctors and technicians that the laboratory brought into being.[2]

Bruno Latour's contention that it is '*in his very scientific work, in the depth of his laboratory*' that '*Pasteur actively modifies the society of his time*' (Latour 1999: 267) points in the same direction. For it was his ability to mobilise the microbes produced in his laboratory that gave Pasteur the ability to reshape society as an example of the processes of translation understood as encompassing 'all the negotiations, intrigues, calculations, acts of persuasion and violence, thanks to which an actor or force takes, or causes to be conferred on itself, authority to speak or act on behalf of another actor or force' (Callon and Latour 1981: 279). By virtue of the microbes he was able to control and interpret, and which were enrolled as actants in and through the practices of the whole corps of socio-medical personnel who invoked them as allies in their strategies for managing the social, Pasteur's scientific activities became directly political:

> If by politics you mean to be the spokesman of the forces you mould society with and of which you are the only credible and legitimate authority, then Pasteur is a fully political man. Indeed, he endows himself with one of the most striking fresh sources of power ever. Who can imagine being the representative of a crowd of invisible, dangerous forces able to strike anywhere and to make a shambles of the present state of society, forces of which he is by definition the only credible interpreter and which only he can control.
>
> (Latour 1999: 268)

The scope for thinking of museums analogously as places in which new forces and realities are constructed, and then mobilised in social programmes by those who are empowered to act as their credible interpreters, is readily perceptible. Museums have served as important sites for the historical production of a range of new entities (like art, community, prehistory, national pasts or international heritage), which, through contrived and carefully monitored 'civic experiments' directed at target populations (the workingman, children, migrants) within the museum space, have been brought to act on the social in varied ways.[3] The role that museums have played in mapping out both social space and orderings of time, in ways that have provided the vectors for programmes of social administration conducted outside the museum, has been just as important, playing a key role in providing the spatial and temporal coordinates within which populations are moved and managed. 'Give me a laboratory and I will raise the world' is the title of one of Latour's articles (Latour 1990). This suggests, as a rough equivalent, 'Give me a museum and I will change society' in view of the museum's capacity, through the studied manipulation of the relations between people and things in custom-built environments, to produce new entities that can be mobilised – both within the museum and outside it – in social and civic programmes of varied kinds.

However, before pursuing this line of thought further, it needs to be acknowledged that both Knorr-Cetina and Latour have been taken to task for working with a more generous interpretation of the idea of the laboratory than others think is warranted.

This is the view of Ian Hacking who has cautioned against seeing too much overlap between laboratories and more open network systems (Hacking 1992). Moreover, whereas Latour includes collections and archives in his definition of laboratories,[4] Hacking explicitly excludes the classificatory and historical sciences from his definition, which stresses instead the ability of the laboratory sciences to interfere directly with the object of study. 'The laboratory sciences', he writes, 'use apparatus in isolation to interfere with the course of that aspect of nature that is under study, the end in view being an increase in knowledge, understanding, and control of a general or generalisable sort' (Hacking 1992a: 33). This means, Hacking concludes, that while museums may undoubtedly contain laboratories in their basements, they cannot be so considered in their archival and classificatory functions since these lack, or do not comprise, an apparatus of intervention on the laboratory model.

The point is debatable. John Pickstone has shown how the intellectual operations of a range of early-nineteenth-century sciences were essentially museological in the respect that their comparative and classificatory procedures depended on the ability to make observations across the large bodies of material collected in museums, and to abstract from these the systems of relations between them that their assemblage in collections made visible (Pickstone 1994 and 2000). Nélia Dias also reminds us of the relations between the skull collections of late-nineteenth-century anthropological museums and craniological experiments, and between both of these and the forms of intervention in the social represented by the administration of colonised peoples (Dias 2004: 220–28). However, when these caveats are entered, I suspect that, technically speaking, Hacking is right here. The knowledges that have been most closely associated with the development of museums and that have provided the basis for their curatorial specialisms have been a mix of historical and cultural sciences that, while often drawing on laboratory sciences (through carbon-dating techniques, for example), have typically fabricated the entities they construct by different means (the fieldwork situation and the archaeological dig, for example). Morever, as we shall see in Chapter 5, in cases like that of the Musée de l'Homme, where museums have modelled aspects of their concerns on the laboratory sciences, this has rarely been entirely convincing. Yet the laboratory analogy is still a productive one in drawing attention to the ways in which the museological deployment of such knowledges brings objects together in new configurations, making new realities and relationships both thinkable and perceptible. The crucial point, though, is the occurrence of this within a space that is simultaneously epistemological and civic, for it is this that enables such assemblages and the relationships between persons they enter into in the museum space to constitute an apparatus of intervention in the social.

It will be useful, in developing this argument, to go back to the three aspects of laboratory science identified by Knorr-Cetina: namely, that laboratories do not have to make do with objects as, where or when they 'naturally' occur. For it is true of the museum just as much as it is of the laboratory that it does not need to 'put up with the object as it is'. The museum object is, indeed, always non-identical with itself or with the event (natural, social or cultural) of which is the trace. Its mere placement within a museum frame constitutes a detachment from its 'in itselfness', and one that renders it amenable to successive reconfigurations through variable articulations of its relations to other, similarly constituted objects. It is equally true that in 'bringing objects home', detaching them from where they 'naturally' occur, museums are able to manipulate those objects on their own terms in ways that make new realities perceptible and available for mobilisation in the shaping and reshaping of social relationships. This was, indeed, Hegel's central contention concerning the productivity of the art museum. By severing the connections linking works of art to the conditions of their initial production and reception, the art museum opened up the space for a properly historical cultural politics that would be alert to the possibilities presented by transformations of the relations between cultural artefacts. It constituted a space in which the meanings and functions of artefacts could be made more pliable to the extent that, once placed in a museum, they were no longer limited by their anchorage within an originating social milieu or immanent tradition (see Maleuvre 1999: 21–9). Latour's conception of anthropological and natural history museums as centres of collection in which objects from a range of peripheral locations are brought together is another case in point (Latour 1987: 223–8), one that foregrounds the relationship between collecting expeditions and museums as 'the sites in which all the objects of the world thus mobilised are assembled and contained' (Latour 1999a: 101). Their functioning in this regard has played a pivotal role in the organisation of the socio-temporal coordinates of colonialism as a consequence of the differentiation they established between, on the one hand, the far away and the primitive and, on the other, the close at hand and the modern (Bennett 2004: 19–24).

There are, however, many other examples that might be cited. The *ensemble ecologique* developed by Georges-Henri Rivière at the Musée des Arts et Traditions Populaires in Paris has played a key role in producing, as both a surface of government and a locus for new forms of agency, communities identified in terms of the regional cultural ecologies, or territorially defined ways of life, that such arrangements make visible (see Poulot 1994). The close relations between art museums and art history in producing art as an autonomous entity as the necessary precondition for its (contradictory) mobilisation in civilising programmes, or as a key marker in processes of social differentiation, is another case in point. In all of these cases, museums are a prime example of those processes through which technologies are able to accumulate in themselves powers and capacities derived from the different times, places and agents that are folded into them through what they bring together – powers and capacities that can then be set in motion in new directions (Latour 2002). They are all cases, too, in which the productive power of institutions is made manifest in their ability to fabricate new entities out of the materials they assemble. 'Boeing 747s do not fly, airlines fly': this is how Latour once summarised his contention that only corporate bodies could absorb and regulate the proliferation of mediators through which we and our artefacts have become 'object institutions' (Latour 1999a: 193). If it is true, similarly, that it is art museums and not artists who make art, the perception is one that needs to be extended to the wider range of entities (community, heritage, regional cultures, etc.) that are produced by museums as 'object institutions' *par excellence*.

If such entities are museum fabrications, then, to come to the third aspect of Knorr-Cetina's characterisation of laboratory science, it is through the observation of the effects of the different orderings of the relations between visitors and such entities that museums dispense with 'natural' cycles of occurrence to organise experimental situations, in which contrived and staged encounters between people and objects can be arranged for the purpose of both continuous and comparative study. Museums are, in this regard, one of the most intensively monitored spaces of civic observation that we have, with countless studies drawing on a plethora of quantitative and qualitative techniques (sociological and psychographic visitor profiling, exit interviews, time and motion studies, etc.) to assess and calibrate the museum's precise civic yield in terms of learning outcomes, improved visitor attentiveness, increased accessibility, social cohesion or greater cross-cultural understanding. And it is through the variety of ways in which they thus monitor and assess the outcomes of such 'civic experiments' that museums generate 'civic results' that are portable from one museum to another.

While acknowledging that there are limits to how far the museum/laboratory analogy can and should be taken (and not least because visitors practice their own forms of often quite unpredictable agency within the museum space) there is one further aspect of Knorr-Cetina's approach to laboratories I want to draw on. It is her contention that what she calls the 'epistemic objects', which are produced through the research process in settings like laboratories, have a complex, necessarily unfinished structure that breaks with everyday and habitual relations to objects to generate an ongoing creative intellectual engagement with them. Here is how she puts it:

> The everyday viewpoint, it would seem, looks at objects from the outside as one would look at tools or goods that are ready to hand or to be traded further. These objects have the character of closed boxes. In contrast, objects of knowledge appear to have the capacity to unfold indefinitely. They are more like open drawers filled with folders extending indefinitely into the depth of a dark closet. Since epistemic objects are always in the process of being materially defined, they continually acquire new properties and change the ones they have. But this also means that objects of

knowledge can never be fully attained, that they are, if you wish, never quite themselves.

(Knorr-Cetina 2001: 181)

It is, Knorr-Cetina argues, this open and unfolding, never completed, form of objecthood, one that is always at odds with itself, that produces the epistemic desire that motivates and animates the process of scientific inquiry. One question to be considered, then, is whether and, if so, how the forms of objecthood that are produced by museums are characterised by a similar internal complexity that gives rise to similarly complex and dynamic forms of interiority on the part of the persons who become entangled with them. A second concerns the respects in which, contrarywise, the object regimes of particular types of museum might, as Knorr-Cetina puts it, have the character of 'closed boxes'. It is to this question that I now turn as one that goes to the heart of debates concerning the relations between museums and liberal government.

Cultural objecthood and self/other governance

There are two general aspects of objecthood I want to consider here. The first concerns the respects in which the arrangement of the relations between the individual objects that are assembled together in museums bring into being the more abstract entities – like art, prehistory, community, national heritage – that then subpoena those objects as aspects of the realities and relations they organise. The focus here is thus on the operations of cultural institutions in producing distinctive kinds of objecthood understood as a product of the arrangements of objects that they effect rather than on distinctive kinds of objects: objects classified as natural are just as much caught up in distinctive kinds of cultural objecthood through their inscription in natural history displays as are objects classified as archaeological or as artistic in art museums. Made durable and sustainable by the institutional ordering of the relations between material objects, such regimes of objecthood operate much like the quasi-objects Michel Serres discusses in his account of the role that the stabilisation of objects plays in the constitution of social relations (Serres 1982: 224–34; see also Brown 2002 and Latour 1993: 51–5). Like the tokens in a game that, for Serres, constitute the paradigm case of quasi-objects, such regimes of objecthood are very much 'in play' in the processes through which social collectives of various kinds (whether classed, regionalised, gendered, racialised or sexualised) are organised. These processes are enacted, through the positions that such collectives take up in relation to each other via the quasi-objects that mediate these moves, and counter-moves, of identity formation. Niklas Luhmann recognises this in the case of works of art that he interprets as quasi-objects, in the sense that they maintain their concreteness as objects throughout changing situations while also assuming 'a sufficient amount of variance ... to keep up with changing social constellations' (Luhmann 2000: 47) in view, precisely, of the fact that their significance as art is not a given but derives from their social regulation.

Second, however, Knorr-Cetina's observations on the open structure of 'epistemic objects', and its consequences for the organisation of epistemic desire and the distinctive forms of activity and relations to the self this makes possible, suggest the need to also take account of the ways in which the regimes of objecthood produced by museums are differentiated with regard to the distinctive kinds of work and self-work that these make possible. What kinds of complex inner organisations do objects acquire from their insertion in different regimes of objecthood? What kinds of interiority on the part of the subject do these enable and/or require? What kind of work of self on self do different kinds of cultural objecthood make possible? Or what kinds of closure do they operate? And how are their roles in these regards related to processes of identity formation? Knorr-Cetina's own interests in these issues have been prompted by her concern with the relations between object regimes and the libidinal aspects of scientific inquiry. However, her contention that 'objects understood as continually unfolding structures which combine presence and absence will have to be added to the sociological vocabulary' (Knorr-Cetina 1997: 15) suggests a broader canvass in which the organisation of object regimes associated with specific forms of expertise becomes crucial to 'postsocial' understandings of the organisation of contemporary forms of sociality.

It will be useful, in exploring the relations between these two aspects of cultural objecthood, to consider Donald Preziosi's arguments concerning the ways in which the historicisation of objects in the art museum has put those objects into play in processes of identity formation. The key aspect of the art museum's operations in this regard, he argues, consists in its capacity to make objects 'time factored' such that they 'are assumed to bear within themselves traces of their origins; traces that may be read as windows into particular times, places, and mentalities' (Preziosi 2003: 19). The art museum adds to this a classificatory operation, according to which the place of each individual work is fixed by assigning it an address within 'a universally extensible archive within which every possible object of study might find its place and locus relative to all others' (24). It also adds an evaluative operation in which, once placed in this archive, works of art are ranked relative to one another in terms of their historical 'carrying capacity': that is, the semantic density of the historical information that is coded into them for retrieval. The greater this is, the greater the work of art in view of the greater degree to which, compared with works accorded a lower 'carrying capacity', it can thus be subpoenaed in testimony to the cumulative production of the universal that was the art museum's post-Enlightenment project. It is, Preziosi contends, this quality of the artwork, a product of the art museum, which allows it to be connected to the processes through which social collectives are formed, differentiated and ranked hierarchically in relation to each other. As he puts it:

> The most skilled works of art shall be the widest windows onto the human soul, affording the deepest insights into the mentality of the

maker, and thus the clearest refracted insights into humanness as such. The 'art' of art history is thus simultaneously the instrument of a universalist Enlightenment vision and a means for fabricating qualitative distinctions between individuals and societies.

(Preziosi 1996: 36)

Valuable though these insights are in underlining the relations between the art museum's hierarchical ordering of differences and the formation of ranked social collectives, there are other aspects to the depth structure of the artwork that is produced by the historicising procedures of the art museum that Preziosi's formulations do not quite fathom. Hans Belting (2001) provides a useful point of entry into these in his assessment of the distinctive kind of objecthood these procedures give rise to as each work of art comes to be haunted by the ideal of 'absolute art', which, while motivating artistic practice, also necessarily eludes it. This ideal differs from the 'classical masterpiece' of the eighteenth century when canonicity was more typically a matter of producing an artwork that would conform as closely as possible to the prescribed formulae of the academies. The canonicity of the modern artwork, by contrast, Belting suggests, is tilted forward, constantly pointing, as a step beyond or behind the ends of the developmental series the art museum constructs, to an unachievable ideal of perfection – an 'invisible masterpiece' that remains always hidden and out of reach.[5]

Belting's interest in the form of objecthood produced by the art museum primarily concerns its dynamising consequences for the forms of artistic production associated with the modern art system with its succession of avant-garde movements. While pitching themselves against the prevailing forms of the art museum's canonicity, these succeed only in realising its deep structure by eventually falling into their preordained places as the most recent approximations to, yet still incomplete realisations of, the ideal of absolute art that is the promise of unblemished perfection held out by the 'invisible masterpiece'. This form of objecthood has, however, proved equally consequential in fashioning those historically distinctive forms of interiority through which the kinds of work of self-on-self associated with aesthetic relations to the work of art proceed. For the incompleteness of the artwork has also served as a template for the organisation of a division within the person between the empirical self and an unreachable ideal that motivates an endless process of self-formation as the beholder strives to achieve the ideal, more harmonised, full and balanced self that is represented by the standard of perfection embodied in the idea of the absolute work of art that hovers just behind or beyond the art on display.

This, the central 'civic experiment' of the nineteenth-century art museum, was made possible by the liberation of the lower faculty of the aesthetic from its tutelage to the higher one of reason that had characterised the relations between aesthetic thought – especially that of Christian Wolff – and state reason in the arts policies of the Prussian state. As I look at these questions more fully in Chapter 6, I shall settle for a brief summary here. Wolff's formulations, as Howard Caygill summarises their political effects, 'restricted the scope of culture to the cultivation of the lower, sensible faculty by the higher, rational one' and, thereby, made 'the realisation of perfection and freedom' the responsibility not of 'individual citizens making judgements in civil society, but that of philosopher bureaucrats who judged what was best for the common good' (Caygill 1989: 141). Kant's autonomisation of the aesthetic as an independent form of cognition allowed a reconceptualisation of the space of the art museum as one of self-formation through the acts of judgement – and, via the artwork, of self-judgement and formation – on the part of a free citizenry.

Yet, if this annexes art to the practices of liberal government in the stress these place on the participation of the governed in governing themselves, it would be wrong to conclude that Kant's position has ever entirely eclipsed the earlier, Wolffian orientation. It is, indeed, possible to illuminate significant aspects of the subsequent history of art museums in terms of the different ways in which these two orientations have inscribed artworks in both pro-cesses of social differentiation and those of governance. Art museums have, in the case of twentieth-century practices of connoisseurship for example, formed a part of the processes through which classed collectives have been formed by differentiating those able to enter into the transformative relation to the self that the artwork's conception as an 'invisible masterpiece' makes possible from those who are judged to lack this capacity (McClellan 2003). Equally, though, works of art have also been caught up in 'extension movements' where what is at issue is organising an extended circuit for the distribution and circulation of art objects that will co-opt new constituencies into the transformative relation to the self that engagement with the complex inter-iority of the artwork makes possible (Barlow and Trodd 2000). Except that, very often, the artwork, when it is brought into contact with 'the people', loses that complexity in being refashioned as a more straightforwardly didactic instrument – a closed box in Knorr-Cetina's terms – through which popular tastes and practices of seeing are to be managed and regulated in a more directive fashion.[6]

The relations between art museums and forms of cultural objecthood are, in other words, variable depending on the nature of the civic experiments in which they are engaged and the populations concerned. If the insights that can be generated by looking at museums as laboratories are to be generalised, however, attention needs also to be paid to the distinctive forms of cultural objecthood that have been fabricated by other types of museum. This is something I undertake in *Pasts Beyond Memory* (Bennett 2004) where, by examining how the uses of the historical sciences (archaeology, anthropology, palaeontology and geology) in evolutionary museums contributed to the fab-rication of the new entity of prehistory, I also consider how this produced a new and distinctive kind of objecthood in the form of 'archaeological objects'. These were objects that, whether in natural history displays or in exhibitions of the development of technologies, weapons or ornaments, acquired a new depth structure by being interpreted as summaries of the stages of evolution preceding them. Viewed as the accumulation of earlier phases of development that had been carried forward from the past to be deposited in the object as so many sedimented layers, the archaeological architecture of the objects in evolutionary museums produced a new form of complex interiority in the object domain. Objects in typological displays, for example, constituted both a summary of the earlier stages of development stored up in the objects preceding them as representatives of earlier stages of evolutionary development, and a departure from that accumulated past in registering a new stage of development. Yet, as well as storing the past, each object in such displays also points to the future: not what it once was, it is not yet what it will become – thus introducing a new kind of dynamic historicity into the field of objects.

This found its echo in a parallel construction of the person as similarly an archaeologically stratified entity made up of so many layers of past development folded into the organisation of the self. Operating in the relations between these forms of objecthood and personhood, evolutionary museums functioned as mechanisms for differentiating collectives, primarily racialised ones, in the form of a division between those with thickly and those with only thinly stratified selves: white Europeans in the first case, black 'primitives' in the second. In relation to those with thickly historicised selves, evolutionary museums served as a template for a process of developmental self-fashioning in which the legacy of earlier layers of development was to be sloughed off in order to update and renovate the self so that it would be able to respond appropriately to the imperatives of social evolution. Where this architecture of the self was judged to be absent – as was the case with the construction of the Aborigine as 'primitive', an evolutionary ground zero – the template provided by the archaeological structure of the museum object was deployed differently.

We can see this in the early-twentieth-century programmes of Aboriginal administration that were developed by Baldwin Spencer. Again, since I engage with Spencer more fully, albeit from different perspectives, in both Chapter 4 and Chapter 8, I shall address only the bare essentials here. Cutting his museological teeth by assisting in the arrangement of Pitt Rivers' typological displays when they were moved to the University of Oxford, Spencer subsequently introduced typological principles into the arrangement of Aboriginal artefacts at the National Museum of Victoria in Melbourne during his tenure as its second director. When he later became involved in the administration of Aboriginal affairs, Baldwin introduced the principles of sequence on which typological displays rested into the civilising programme he proposed for Aborigines. This involved forcibly removing 'half-castes' from their communities and moving them through a series of staging houses until, once they had been 'fully developed', they would be able, at the end of the process, to be absorbed into white society. The logic underlying this programme depended on interpreting the racial impurity of 'half-castes' as a sign of developmental possibility in comparison to the utterly flat, undeveloped makeup of the 'full-blood' and consisted in the movement of bodies through social space as if they were so many museum pieces being moved along a continuum of evolutionary development. As such, it aimed at the compulsory introduction of sequence into the Aboriginal population – previously (and, of course, erroneously) judged to lack it – as a necessary prelude to their being accorded, but only as individuals severed from their communities, a capacity for self-government.[7] Once the epidermal-cum-cultural transformation of the Aborigine that this programme envisaged had been managed to the point of giving rise to an archaeological splitting of the Aboriginal self into a division between its primitive and archaic layers, on the one hand, and its civilised and modern ones, on the other, then so the Aboriginal – as, now, an individualised persona – would be able to assume direct and personal responsibility for his or her own evolutionary and civic self-fashioning.

This is a telling example of the respects in which, to recall Knorr-Cetina's formulations, museums, like laboratories, 'create new configurations of objects that they match with an appropriately altered social order'. For it illustrates how distinctive configurations of the relations between things produced by the deployment of particular knowledges within museums actively shape the contours of the social within which, once they are mobilised by agents outside the museum, those new realities and relations become active agents in specific programmes of social management. Dominique Poulot's sharp observations regarding the redistribution of national heritage associated with the development of ecomuseums point to a similar set of processes. By breaking the national heritage down into discrete environmental cultures, Poulot argues, the ecomuseum has formed a part of the 'regionalisation of the social', allowing its organisation to be conceptualised as a set of relations between regionally defined communities (Poulot 1994). In being pitted against earlier French statist conceptions of the museum as a civic technology acting on citizens who are placed in direct, unmediated and identical relations to the state, the ecomuseum produces the territorially defined community as a key point of identity formation. It does so, moreover, by reversing the operation of the universal survey museum, reattaching objects to the specific regional cultural systems that the latter had detached them from, and, thereby producing, in the notion of a regional cultural ecology, a new surface of civic management – a space in which identities can be caught and nurtured in spatially defined programmes of cultural or community development. This, in its turn, is connected to the specific forms of objecthood that the ecomuseum fashions by resocialising ordinary and familiar objects. By placing these in the context of what might be an environmentalist, geographical, folkloric or sociological knowledge of the operative principles connecting them together in a distinctive cultural ensemble, the ecomuseum enlists such objects for new processes of identity formation that depend on the active acquisition of new forms of self-knowledge rather than, as Hegel had feared, simply confirming existing identities as organic quasi-vegetal entities rooted in the local soil.

A final and related example is given by Bill Brown's discussion of the regionalisation of the artefactual field that was carried out in late-nineteenth-century American anthropology. The key figure here was Franz Boas whose displays, focused on life-group exhibits in specific geographically defined tribal and environmental settings, served, as Brown puts it, 'to "regionalise" anthropology, and to displace artefacts from their traditional place within a taxonomic and evolutionary scheme' (Brown 2003: 88). Brown's concern is with the relations between this regionalised object-based epistemology and parallel tendencies in American letters, especially the writing of Sarah Orne Jewett, seeing these convergences as parts of new ways of operationalisating regionalism as a framework for thought and action. It is relevant in this regard to note the respects in which Boas' work contributed to a distinctive kind of 'regionalisation of the social' that, in the post-Indian Wars context, replaced the temporal coordinates that had been proposed by evolutionists like Spencer for the management of indigenous peoples with a new spatialised conception of indigenous cultures as regionally distributed, holistic ways of life (Hinsley 1981).

In summary, then, this conception of museums as 'civic laboratories' calls attention to the way in which the ordering and classificatory practices that are conducted within museums – ordering practices that are conducted across their research, exhibition and archive functions – give rise to distinctive forms of objecthood through which museum artefacts are equipped with specific capacities. It is through these capacities that they are enlisted by, and in turn enlist, other agents – community workers, for example – as parts of distinctive collectives that operate on the social in varied ways. Such collectives do not, however, act directly on the social as a given but through the working surfaces on the social that they organise as the mediatory interfaces through which their engagements with the social are enacted.

Re-socialising objects, diversifying the social

I turn now to the directions in which these remarks point when they are brought to bear on current attempts to redeploy museum collections for new civic purposes by re-socialising the objects that are contained within them so that they might function as the operators of new kinds of action on the social. These have centred, I suggested in the previous chapter, on the promotion of forms of inter-group conduct governed by the diagram of tolerance. The most important of these tendencies consist in the now more-or-less ubiquitous concern to refashion museums as 'differencing machines' that, in varied ways, are intended to ameliorate conflicted racialised differences. What different kinds of cultural objecthood are produced by the reconfiguration of the relations between objects, and between objects and persons, within museums where such concerns predominate? What role do these play in putatively reshaping the social by being mobilised as parts of civic programmes that aim to act on the relations between ethnically differentiated communities as opposed to those between hierarchically ranked social classes?

It will help in answering questions of this kind to identify how the approach developed in the foregoing discussion differs from those forms of discourse analysis that – quite contrary to those advocated by Foucault, which, as Frow reminds us, treat discourses as complex assemblages of texts, rules, bodies, objects, architectures, etc. (Frow 2004: 356) – convert the museum into a text that is to be analysed to reveal its ideological effects in occluding the real nature of the social relations it represents.[8] These typically exhibit three main shortcomings. The first concerns the lack of any clear attention to the distinctive forms of objecthood associated with different kinds of museum. Dissolving objects too readily into texts in order, thereby, to make museum arrangements readable as ideologies, such approaches fail to grapple adequately with the different and specific kinds of qualities objects acquire as a result of the ways in which the relations between them are configured in different museum practices. Second, and as a consequence, they fail to deal adequately with the distinctive operations, procedures and manipulations through which different knowledges – art history, anthropology, natural history – fabricate new entities through the new alignments of the relations between objects that they establish. They deal with such questions largely abstractly by positing homologies between the intellectual structure of particular knowledges and museum arrangements, paying little attention to the whole set of technical procedures (from accessioning, classification, conservation, etc.) through which objects are actually manipulated and managed. And third, they pay little attention to the distinctive relations to the self and ways of working on it that are made possible by different forms of objecthood.

The kind of displacements of these approaches that my comments point towards echo those advocated by Alfred Gell in his concern to develop an action-centred approach to art in which art is viewed as 'a system of action, intended to change the world rather than to encode symbolic propositions about it' with the consequence that it is 'preoccupied with the practical mediatory role of art objects in the social process, rather than with the interpretation of objects "as if" they were texts' (Gell 1998: 6). The difference, however, and it is one that Gell points out himself, is that when such perspectives are applied to the distributed relations between objects and persons associated with Western art institutions, it is the action of art objects in the supra-biographical relations between classes, or castes, or status groups, or communities that has to be attended to rather than, as Gell's own concern, their role in mediating more personalised forms of social interaction. And this means, as he puts it, attending to the institutional processes through which some objects come to be 'enfranchised' as art (Gell 1998: 12). In seeking both to apply this perspective to art museums and to generalise it to the processes through which other distinctive kinds of objecthood are produced within Western collecting institutions, the methods of science studies are helpful precisely because of the centrality they accord to those processes that approaches to museums as texts neglect. They bring with them the kind of attention to technical procedures through which the realities that science works with take on a phenomenal form by virtue of their construction through material techniques. Such attention is necessary if the operation of museums as epistemological-cum-civic technologies, working on and with objects in distinctive ways, is to be understood in adequately specific terms.

At the same time, we should be wary of the temptation to approach these questions in terms of epochal divisions of the kind implied by Foucault's historical sequence of epistemes. This is so for three related reasons. First, the ways in which the relations between objects are configured and, accordingly, the agency that is attributed to them, is always provisional. The fixity into which they are ordered is always a loose and pliable one. Grahame Thompson puts this well in his account of the operation of 'immutable mobiles' in actor–network theory. While always the same, such objects – and the museum object is a classic example – take on different values and functions when moved from one set of configurations to another. 'Whilst "fixed" in one sense', Thompson says, 'these are also made "mobile", by being arranged and reconfigured through the network of places and agencies to which they are attached and through which they operate; they have the combined properties of mobility, stability, and combinability' (Thompson 2003: 73).[9] Second, a point I take from Latour's account of the ways in which technologies fold into and accumulate within themselves powers and capacities derived from different times and places (Latour 2002), objects carry with them a part of the operative logic characterising earlier aspects of their history as they are relocated into reconfigured networks. Their organisation in this respect is, Latour argues, always archaeological as aspects of their earlier use and inscriptions are sedimented within them. Third, this process of enrolling objects in networks always has multiple, and often contending, dimensions at any one time just as it lacks a single point of origin or definite finality.

These considerations caution against ruptural accounts, such as those based on Foucault's notion of epistemes, in which museum objects are said to be disconnected from one configuration to be inscribed in another governed by entirely different epistemological principles.[10] This is not to dispute the usefulness of such accounts in identifying significant shifts in the permissible forms of combinability of objects that have proved relatively durable and widespread. It is, though, to caution against the view that such shifts entirely cancel the earlier operative logic of the objects they enrol, or that they are only enrolled in single and stable configurations at any one point in time (see, on this, Daston and Galison 2010). This, of course, is precisely what we find in the current flux and fluidity of museum practices: not a simple transition from one episteme to another, but a profusion of different ways of rearranging the relations between objects and persons in the museum environment and of enrolling these in social and civic programmes of varied kinds.[11]

Yet it is here, perhaps, that the laboratory analogy begins to break down. For – and this is where Hacking's (1992) objections ring true – the laboratory situation usually involves a more singular and authoritative manipulation of the relationships between objects than has been true of museums over the last quarter of a century or so. The challenges to the classificatory procedures of the cultural and historical sciences that are the mainstay of Western curatorial practices have come from a range of quarters: from postcolonial theory, and indigenous critiques and counter-knowledges, for example. The divisions that these have occasioned within Western systems of thought closely associated with museums, especially archaeology and anthropology, mean that there is now a much greater tug-of-war between competing knowledges regarding the arrangement of the relations between objects and persons within the museum space. There is little disagreement – at least, little public disagreement – with the view that such relations should be reordered with a view to reconfiguring the social in more culturally plural ways. All the same, this shared commitment belies a real variety of practice and effect as museums are variously conceptualised as contact zones, as spaces for dialogic encounters between cultures, as instruments for the promotion of cultural tolerance, or as means for promoting and managing the identities of differentiated communities.[12]

Rather than developing this point abstractly, a brief comparison of the *Living and Dying* exhibition at the British Museum with contemporary rearrangements of Aboriginal materials in Australian museums will help to identify some of the issues posed by these debates and practices. Opened in 2003 as a part of the British Museum's 250th anniversary celebrations, *Living and Dying* involved the re-socialisation of objects from both the British Museum's own collections and those of the Wellcome Trust. Lisant Bolton, the curator of the Museum's Pacific and Australian collections, has contrasted her experience as the lead curator of *Living and Dying* with her earlier work at the Australian Museum and the Vanuatu Cultural Centre. Her account usefully foregrounds some of the tensions concerning the relations between different curatorial practices, the ways in which they are institutionally authorised, and the implications of these considerations for the ways in which museums construct and fashion the social they seek to act on (Bolton 2003). In the Australian Museum, Bolton argues, the authority exercised by Indigenous Australians over how indigenous cultural materials are presented results in forms of cultural advocacy that expose and critique the historical particularity of the earlier colonial frameworks in which such materials had been exhibited.

A more recent example, and one I am more directly familiar with, is that provided by Bunjilaka, the Aboriginal Centre at the Melbourne Museum, where Aboriginal curation and extensive community consultation has resulted in a re-socialisation of the Aboriginal weaponry that had been a part of the evolutionary exhibitions introduced into the National Museum of Victoria by Baldwin Spencer in the early twentieth century. Displayed in a vitrine alongside a reconstruction of Spencer – so that the collector of Aboriginal culture is collected alongside his collections in an Aboriginal framing of both – these artefacts are wrenched from the evolutionary time in which they had originally been installed to open up a new, indigenously marked time in which the forms of hunting and collecting that characterised anthropological practice along the colonial frontier are depicted as archaic. Equally important, Bunjilaka provides a space that is both in the Melbourne Museum and not in it, of it and not of it, to the degree that it is marked out as a semi-separate space in terms of its location (to one side of the museum), the community-auspiced history of its curation, the Aboriginal guides who mediate the visitor's relations to the Centre, and its inclusion of a meeting space for the conduct of Aboriginal affairs. The consequence is a form of cultural objecthood that establishes a tutelary relation in which an Aboriginal framing of objects in the present seeks to detach them – and the visitor – from their past inscriptions, thus mobilising the museum's capacity as a civic and reformatory apparatus in distinctive ways.

The contrast Bolton draws between such strongly marked re-socialisations of objects and the British Museum is not, however, one in which the ethnographic authority of the curator prevails over that of indigenous knowledge so much as one in which the institutional 'voice' of the museum prevails over both. Conceived as a cross-cultural exhibition on the theme of health and well-being, *Living and Dying* examines, in the words of the exhibition's summary description, 'how people everywhere deal with the tough realities of life, the challenges we all share but for which there are many different responses' – sickness, trouble, sorrow, loss, bereavement and death. The main organising principle for the exhibition echoes the concerns of contemporary anthropological theory (Bolton studied for her PhD with Marilyn Strathern) by exhibiting objects as mediators in complex and varied sets of relations: those between Native Americans and animals, represented as non-human persons, in the vitrine focused on the theme 'Respecting animals'; and the relations between human persons in the Pacific Islands in the vitrine focused on the theme 'Sustaining each other', for example. Eschewing any normative framework, the exhibition is presented as a testimony to the insights into the variable responses to shared human problems that can be gained from bringing together under one roof so many objects from many different cultures and periods. This results in a distinctive, although not uncommon, form of objecthood in which the specific, culturally variable meanings of any particular set of object-mediated relations are eclipsed in being subpoenaed as stand-ins for anthropologically constant, universally shared human concerns. The result is an exhibition of a set of what are largely 'disconnected diversities' – disconnected from each other as well as from any particular histories connecting them to each other in either allied or hostile relations – as a testimony to the creative ordering capacity of human beings as evidenced by the varied ways they respond to, and make sense of, death, pain and suffering. In *Mythologies*, Roland Barthes discussed a photographic exhibition held in Paris under the title *The Great Family of Man*, which, much like *Living and Dying*, aimed 'to show the universality of human actions in the daily life of all the countries of the world: birth, death, work, knowledge, play' (Barthes 1972: 100). The effect of this, Barthes argued, was to magically produce unity out of pluralism: the diversity of different ways of life does not belie, and cannot eclipse, 'the existence of a common mould' (Barthes 1972: 100). The effect of *Living and Dying* is not quite the same: rather than invoking a universal humanness, it reconfigures earlier hierarchical forms of difference into an abstract form of 'side-by-sideism' through which the social is mapped as a set of equivalent differences.

The significance of this, however, is fully evident only when *Living and Dying* is viewed in the context of its juxtaposition with the *Enlightenment* exhibition. Opened at roughly the same time, these two exhibitions – both of them permanent – are also adjacent to one another, occupying connected sides of the Great Court, and both have featured strongly in the British Museum's institutional discourse in seeking to fashion a new role for itself as a universal survey museum. At one level, the aim of the *Enlightenment* exhibition is a self-consciously relativising one: in seeking to rediscover how the world was intellectually ordered at the time of the British Museum's foundation, the Enlightenment's claims to universality are discrowned by being revealed in all their historical particularity and peculiarity. Yet this message of historical difference, like that of cultural difference suggested by the *Living and Dying* exhibition, depends on an affirmation of the continuing value of the universal survey museum for its ability to bring together people and things from all places and times for the purpose of exploring relations of similarity and difference. The relations between the two exhibitions thus comprise the means by which the universal survey museum is detached from the particular normatively weighted concept of universality associated with the Enlightenment to be, in the words of its director, 'reinvented' as a machinery for exploring relations of sameness and difference (MacGregor 2003: 7) where this means laying out the social as a set of equivalent differences to be tolerated as equally valuable.[13]

And this, in turn, is a part of the institutional script through which the British Museum resists calls for the dispersal of its collections through repatriation, for example. The *Living and Dying* exhibition's place in this scheme of things is thus signalled by its location in what the British Museum describes as its World Cultures galleries exploring the many ways in which different cultures are shaped by their attempts to make sense of 'a common experience of what it means to be human'. The resonances of this are echoed in the wall plaque acknowledging the sponsorship of the Wellcome Trust. Designed as a tribute to Sir Henry Wellcome, it suggests a connection between the exhibition and Wellcome's role as a collector, particularly his unrealised aspiration to establish a Museum of Man – passing in tactful silence over the colonial and evolutionary ordering of the Hall of Primitive Medicine in the Historical Medical Museum that Wellcome *did* establish (Skinner 1986).

Limiting culture

In his account of why critique has run out of steam, Latour groups sociology and cultural studies together, castigating both for the terrible fate they have inflicted on objects (Latour 2004: 165). His objections to sociology have been developed at some length and have now been widely rehearsed in the literature. The rationality of the nature/society distinction is challenged as the simultaneously enabling/disabling fiction of the 'modern settlement'. The division of society into different levels – micro and macro – is criticised as inhibiting understanding of the mechanisms through which particular societies are made up of actor-networks of human and non-human agents, which form chains of connected relations and actions rather than separate levels. Invocations of society as an invisible totality or underlying structure capable of explaining observable actions and relations are interpreted as nothing but a power play on the part of sociologists, from Comte to Bourdieu, in their attempts to lord it over other disciplines (Latour 1993, 2002). As we have seen, however, Latour has paid less attention to the concept of culture, sometimes dismissing it alongside the concepts of nature and society while, at other times, hinting at alternative accounts. The implication, although with some hesitation in his more recent work (see Knox *et al.* 2005), is that there are only actor-networks of humans and nonhumans subject to variable and impermanent inscriptions, translations, articulations and enrolments; and that there are no independently existing grounds (whether of nature, culture or society) outside of these and their effects that can be invoked to explain their operations. 'All that one can do,' as Andrew Pickering summarises the position, 'is register the visible and specific intertwinings of the human and the non-human. But this is enough; what more could one want or need?' (Pickering 2001: 167).

Foucauldian conceptions of the social are, of course, different from sociological ones; indeed, they frame these as a part of a broader account of the emergence of modern forms of governmentality and their production of both the economy and society as autonomous realms, differentiated surfaces of government constructed through the application of new forms of description, classification and enumeration. In his explorations of this analytical territory, Timothy Mitchell (2002) draws freely on the vocabulary of actor–network theory to account for the processes through which national economies are assembled out of a variety of human and nonhuman forces and agencies, constantly stressing their role in the making of new realities and processes of production and exchange. It is in this light, and for the same reasons, that I have suggested – contra Pickering – that the concept of culture retains a similar validity provided that we interpret it as a historically fabricated – in the sense of 'materially made' not 'invented' – set of entities. I have also indicated how, drawing on Latour's later work, the historically circumscribed differentiations of culture/society/economy/nature might be accounted for as different public orderings of the intertwining of human and non-human forces and agencies.

To sustain this view, however, requires that we place limits on the concept of culture and its uses. This is not merely a matter of stressing its historical specificity. It also requires that we register a certain distance in relation to two of the key defining moves of cultural studies. The first of these, as Francis Mulhern describes it, consists in '*a radical expansion of the corpus*' so as to include the role of the symbolic in everyday life in an expanded definition of culture, while the second consists in 'the *unification* and *procedural equalisation* of the field of inquiry' (Mulhern 2000: 78) that this expanded understanding of culture produces. The difficulty here lies in the second proposition, which does two things: first, it asserts that all kinds of culture, whether 'high' or 'low', are equally important and worthy of study; and second, it asserts that similar methods of analysis can be applied in studying all forms of culture and their role in the organisation of social relationships. While not quarrelling with the first of these contentions, the second is patently not true since, first, it occludes the respects in which the forms of cultural objecthood I have been concerned with are the products of distinctive processes of fabrication, involving specific forms of expertise in specific settings, and second, it forecloses on the possibility of applying specific methods to these in order to disclose how they operate on the customs, beliefs, habits, traditions, ways of life, character systems, etc., which comprise the surfaces of the social to which they are applied in programmes of social regulation and management. As we have seen, many anthropologists – Clifford Geertz and Adam Kuper, for example – have questioned the value of the anthropological extension of the concept of culture insisting, instead, on the value of distinguishing customs, habits, beliefs, etc., from art, literature, etc., rather than bundling them all up into one omnibus concept. The purposes of social and historical cultural analysis will similarly be better served by a more careful differentiation of the varied relations, processes and practices that the extended concept of culture yokes together.

Part 2
Anthropological assemblages

Inter-text 1

My concern in Part 1 was principally to set out my stall by arguing for a view of culture that draws attention to the relations between a particular set of knowledge practices, their inscriptions in cultural assemblages and their roles in acting on varied kinds of conduct as parts of governmental processes. While hinting at the broader applicability of this argument, I have drawn my main examples from the material and governmental entanglements of aesthetics and anthropology. I now, in Part 2, focus on the latter by exploring the relations that were forged between anthropological fieldwork and museum practices in case studies of two of the examples I have already referred to: Melbourne's National Museum of Victoria and the Musée de l'Homme in Paris. I examine these with a view to illustrating their functioning as cultural assemblages in which anthropological expertise is put into action via the mechanisms of both public and milieus. The two studies also illustrate the significance of anthropology's relations to different governmental rationalities as exemplified by its role, in the first case, in a settler colony at a crucial moment in its separation from an imperial formation to establish an independent national governmental domain and, in the case of the Musée de l'Homme, its role at an equally crucial moment in the establishment of a French imperial governmental domain. In addressing these issues, I also further amplify the respects in which the processes of 'making culture' and 'changing society' operate through the different 'working surfaces on the social' that are produced by the deployment of different traditions of anthropological expertise.

Anthropological assemblings

4 Making and mobilising worlds
Assembling and governing the Other

Were the dream-times an invention of Western anthropology or an autochthonous aspect of Aboriginal cosmology? Patrick Wolfe, in arguing the former, attributes the primary responsibility for its fabrication to Baldwin Spencer and Frank Gillen.[1] The term was first used by Gillen in the memoir he contributed to the report of the 1894 Horn Expedition into Central Australia. Discussing Arunta[2] accounts of the origins of fire, Gillen reported that this was believed to have been acquired by ancestors 'in the distant past (*ūlchurringa*), which really means in the dream-times' (Gillen, cited in Wolfe 1991: 200). In his introduction to the report of the Horn Expedition, which he edited, Spencer, transforming Gillen's spelling into *alcheringa*, also widened the term's reference by describing it as a more general system of morality. Used again in their 1904 book *The Northern Tribes of Central Australia*, the concept was subsequently extended beyond the Arunta to encompass other tribes and, in a further step, to function as a pan-Australian marker of a generalised Aboriginal culture. Wolfe interprets this extension of the term as being ideologically motivated rather than resting on ethnographic evidence. Tracing the history of the associations that, in the natural history of the Comte de Buffon, had connected the inability to distinguish dreams from reality to animality and that subsequently, in Darwin's work, were interpreted in evolutionary terms as a marker of the distinction between primitive and civilised humans, his contention is that the anthropological invention of 'the Dreaming complex' was 'the culmination of a historical discourse which subordinated dreaming savages to the level of animal nature' (Wolfe 1991: 206).

He attributes particular significance to the differences between Spencer's position and that of the Lutheran missionary Carl Strehlow who, in the same year as the Horn Expedition, had been appointed to supervise the Hermannsburg Mission in Central Australia. Questioning Gillen's translation and his orthography, Strehlow disputed the contention that the word *Altjira* referred to a distant past. 'The native', as he put it, 'knows nothing of "dreamtime" as a designation of a certain period of their history' (Strehlow, cited in Hill 2002: 141). Interpreting it, rather, as a reference to a god, he attributed a religious significance to it that provided the basis on which a programme of salvation evangelism could be built. By construing the *alcheringa* as purely a moral code, and by insisting on the evolutionary inscription of the dream-times as primitive, Spencer denied the dream-times the religious potential that Strehlow's salvation evangelism needed in order to find something within Aboriginal culture that it might latch on to. It was better, in Spencer's view, to leave the dream-times intact especially since, from an evolutionary perspective, attempts at the spiritual-cum-cultural uplifting of the Aborigine were doomed to failure as the inexorable laws of racial competition meant that extinction was the unavoidable destiny of the Aboriginal race. When passing the Hermannsburg Mission a few months prior to Strehlow's appointment there, Spencer had recommended that it should be closed or run by the government. He viewed the missionary endeavour as not only hopeless in attempting to teach the Arunta abstract ideas they were *'utterly incapable of grasping'* (Spencer, cited in Hill 2002: 58) but potentially dangerous since it undermined their faith in the precepts handed down by their elders and thus threatened to undo colonial forms of rule that worked through the reinforcement of customary morality.

The point at issue between Spencer and Strehlow was thus clearly by no means a purely ethnographic one. Their different interpretations of the dream-times were bound up with, and functioned as operative parts of, different strategies of colonial rule that worked through the organisation of different 'transactional realities' – or 'working surfaces' – mediating the relations between white and Aboriginal populations in ways that would allow the latter to be acted on differently: absorbed into a Christian civilisational programme in the case of Strehlow, or, at this stage in his career, having their passage into extinction soothed via a benign colonial administration that managed the milieus governing the relations between racially defined bodies in the case of Spencer.[3]

I borrow the concept of 'transactional realities' from Foucault who, as discussed in Chapter 2, uses it to refer to a series of historical realities – civil society, sexuality and madness are among the examples he cites – which 'although they have not always existed are nonetheless real, are born precisely from the interplay of relations of power and everything which constantly eludes them, at the interface, so to speak, of governors and governed' (Foucault 2008: 297). Foucault's concern is with the roles played by such transactional realities as parts of 'liberal governmentality' (296). However, the same principles apply – to generalise the issue from the specific example of the dream-times – to the role that different versions of Aboriginal culture and its constitutive elements have played in organising the complex 'transactional realities' across which the agency and counter-agency of both colonised and coloniser have been enacted in the development of, and resistance to, the varied forms of 'indigenous governmentality' that have accompanied the twentieth-century project of Australian state formation.[4]

It is from this perspective that, in this chapter, I consider the role played by Spencer's photographic and museum practices in producing and circulating Aboriginal culture as parts of new socio-material arrangements for the management of Aboriginal bodies in the context of the 'Aboriginal domain'

(Rowse 1992) that increasingly mediated the relations between white and black Australia in the early twentieth century. I shall seek some accommodation between the concerns of governmentality theory and assemblage theory in doing so, outlining the respects in which this key period in the 'museumification' of Aboriginal culture constituted, in Deleuze and Guattari's terms, a form of colonial capture that inscribed Aborigines in a new scientific–governmental assemblage (Deleuze and Guattari 1988: 435; see also Patton 2000: 122–31).

From field to museum and back again

Shortly after returning from the Horn expedition Spencer was appointed to the Board of Trustees of the Public Library, Museums and National Gallery of Victoria, and, in 1899, was appointed to the directorship of the museum. Thereafter, the various expeditions he conducted into Central Australia, building on the partnership he had established with Gillen, were undertaken with a view to bringing back items of Aboriginal material culture, and film and sound recordings, to add to the National Museum of Victoria's collections; and to writing scientific books and articles, and delivering more popular public lectures, based on their field notes, films, photographs and the artefacts assembled in the museum's collections. When later introducing his display of the ethnological collection at the National Museum of Victoria, first arranged in 1901, Spencer stressed the museum's role in mediating the relations between his and Gillen's fieldwork and the texts through which the results of that fieldwork were more broadly disseminated.

> Every specimen figured in *The Native Tribes of Central Australia, The Northern Tribes of Central Australia, Across Australia,* and *The Native Tribes of the Northern Territory* is in the Museum collection, together with the whole of the material, including negatives and phonographic records secured by Mr Gillen and myself during the progress of our work.
> (Spencer 1922: 8)

His actions in this regard are usefully viewed through the prism of Bruno Latour's reminder that the chief problem facing fieldwork expeditions is how to render the worlds they explore mobile so that those who take part in them might take something back with them that they will be able to mobilise within the world that sent them and to which they return. It's worth quoting the relevant passage in full:

> If you wish to go out of *your* way and come back heavily equipped so as to force others to go out of *their* ways, the main problem to solve is that of *mobilisation*. You have to go and come back *with* the 'things' if your moves are not to be wasted. But the 'things' have to be able to withstand the return trip without withering away. Further requirements: the 'things' you gathered and displaced have to be presentable all at once to those

you want to convince and who did not go there. In sum, you have to invent objects which have the properties of being *mobile* but also *immutable, presentable, readable* and *combinable* with one another.

(Latour 1990: 26)

And in *Pandora's Hope*, Latour notes how, in many of the sciences, the relations between expeditions and museums played a key role in the processes of mobilising worlds through which 'nonhumans are progressively loaded into discourse', citing the galleries of the Muséum national d'Histoire naturelle and the collections of the Musée de l'Homme as places in which 'all the objects of the world are thus mobilised and assembled' (Latour 1999: 99–101). Or if, in another Latourian formulation, one wants to 'muster on the spot the largest number of well aligned and faithful allies' (Latour 1990: 23) then, for leaders of scientific expeditions, a museum is a very good place in which to put the 'silent witnesses' they bring back with them.

The relations between Spencer's expeditions and the collections he assembled at the National Museum of Victoria are of a fourfold significance from this point of view. First, they formed part of a key moment in the development of the fieldwork phase in anthropology that significantly reorganised the relations between a range of agents compared to those that had characterised the earlier phase of 'armchair' anthropology.[5] While the latter was also dependent on distant worlds being made mobile and assembled in European museums, those collections were usually assembled via networks of commercial or amateur collectors either plying their wares or acting on instructions. Museums in colonial contexts typically acted as intermediaries for the transit of objects from local sites of collection to metropolitan institutions for interpretation and synthesis by anthropologists who had stayed at home knowing that the world would come to them. Spencer's work ranks in significance with the other major expeditions of roughly the same period, through which anthropologists working in European or American museums mounted scientific expeditions in order to go away and collect the world for themselves, in order that they might return with a more systematic set of immutable mobiles.[6] This also led, as Morphy puts it, to 'the combining of theorist and ethnographer in the same person' (Morphy 1996a: 138) with the consequence that the processes through which other cultures were collected, made mobile, gathered and assembled, exhibited, and translated into inscriptions were more closely integrated. Collecting worlds and mobilising them in distant settings were activities undertaken by the same agents.[7]

The Horn expedition, Philip Jones argues, was the first scientific expedition within Australia that (i) was charged to accord significant attention to the collection of Aboriginal artefacts, (ii) actually did so, (iii) returned those collections to an Australian museum (the South Australian Museum) where the work of scientific interpretation – previously done overseas – was done by a curator, Edward Stirling, who had also been the leader of the expedition team, and (iv) which resulted in that museum providing the first institutional setting for the public exhibition of those artefacts (Jones 1996). As well as continuing this tradition, Spencer played a key role in staunching the flow of Aboriginal objects to Europe and America by supporting the introduction of export licensing requirements, and he contributed considerably to their local accumulation when he donated his own collections to the National Museum of Victoria in 1916 (Lydon 2005: 17). The ethnographic holdings of the museum increased from 1,200 to over 36,000 during Spencer's period as director (Mulvaney and Calaby 1985: 252). Spencer was thus able to install the immutable mobiles he accumulated at the National Museum of Victoria in two sets of relations at once. The *in situ* presence of the collections in the museum gave him a large number of 'well aligned and faithful allies' that he could 'muster on the spot' in the context of specifically national political and administrative fields. These collections also provided the material basis and warrant for the inscriptions through which, in journal articles and books – and the copies of original artefacts he continued to send to European museums – the data that he and Gillen had gathered were recombined in accounts of Aboriginal culture, which had considerable influence on the development of European anthropology and, in serving as the main source data for Durkheim's account of the elementary forms of religious life, sociology too.[8] The international standing that thus accrued to Spencer had significant consequences for his position and authority in the Australian scientific field.

Third, as the first Australian expedition to involve the use of film and sound recording, Spencer and Gillen's first encounter with the Arunta of Central Australia contributed to a significant shift in the organisation of ethnographic authority. According to Wolfe (1999), the phase of Victorian anthropology represented by Edward Tylor's work had been primarily concerned to construct and defend anthropology's claims to disciplinary specialism and authority against those of the materially based discipline of archaeology and the text-based language disciplines, particularly philology. It had done so by proposing that anthropology was able to produce a distinctive knowledge of the prehistoric past by reading its traces as these had survived, not in texts or objects, but in ritual forms of behaviour carried into the present through the habitual mechanism of repetition. With the increasing use of film, photography and sound recording to capture traces of presently existing cultures, however, ethnographic authority became less concerned to define its specificity in relation to the claims of the historical or language disciplines and focused instead on countering rival claims to have gone there, into the field, and to have observed, recorded and collected.[9] It was the colonial administrator, the journalist and the tourist – not the archaeologist or philologist – whose rival productions of the Other had to be disqualified. The prolonged fieldwork of the anthropologist often proved critical here – as it did with Spencer and Gillen's film and photographs of Arunta corroboree – in validating the authority of the ethnographer as having so penetrated the culture of the Other as to have gained admission to secret events (Cantrill and Cantrill 1982).

Fourth and finally, and by way of connecting back to my starting point, if all of these conditions made it possible for Spencer to become an actor on the national stage, it is because they made it possible for him to make and to mobilise a specific production of Aboriginal culture. By bringing things together from diverse locations, combining these in new ways, simplifying and condensing them by subjecting them to further processes of inscription, Spencer produced something that had not existed before: Aboriginal culture not as a set of autochthonous realities that preceded his inquiries but as a new pan-tribal and pan-national surface of connection between white and black Australia, a working surface that organised new sets of governmental and administrative interfaces through which the former might act on the latter. His activities are, of course, by no means unique in this regard. The diverse forms in which photographs, recordings, tools, weapons, sacred objects, films and artworks referenced as 'Aboriginal' have been assembled, recombined and rearranged in relation to each other in different productions of Aboriginal culture have, both before Spencer and since, had significant implications for both different modes of governmental action on Aboriginal communities and Aboriginal counter-actions to these.[10] As Philip Jones notes, however, it was not until the first two decades of the twentieth century that Australian museums installed permanent exhibitions of Aboriginal culture, seeing in this a form of symbolic capture and display of the Other that was closely connected to the process of national state formation following the Federation of Australia in 1901 (Jones 2007: 228, 327–8). Spencer's role was a particularly significant one in this respect, providing a key point of connection between the anthropology/museum nexus and the development of what Tim Rowse has called 'the Aboriginal domain': that is, the domain of indigenous institutions, practices, customs, languages, etc., which, by organising a zone of interaction, provided the means through which different Aboriginal communities were absorbed into the colonial state (Rowse 1992). Rowse identifies the introduction of rationing in 1894 – the year of the Horn expedition – as a key moment in the development of this domain. While by no means replacing the sheer exertion of colonial power and racial violence that had characterised frontier relations in the 1870s and 1880s, with episodic massacres continuing into the 1920s (Rowse 1998), the introduction of rationing to compensate for indigenous food supplies displaced by cattle grazing constituted a significant shift in balance towards the regulation of food supplies as the primary means of control over a dispossessed population (Rowse 1996).[11]

Spencer's role in the Horn expedition involved his participation in this rationing system and, as Rowse records, he was uncomfortable with aspects of it as a form of giving, which, lacking transparency and reciprocity, secured control through an obscured gift relationship (food given in exchange for compliance). Rationing was, however, the basis for the development of a new mode of government that regularised (unequal) exchanges across the colonial frontier and, in place of racial violence, substituted an administrative infrastructure for managing these exchanges. Spencer was to become directly involved in this for a short period when he was responsible for the administration of Aboriginal affairs in the Northern Territory, and he contributed to its development more generally through the various advisory roles he was called on to play.

In discussing the relations between scientific fields and the contexts, such as museums and laboratories, in which specimens from such fields are brought together and subjected to new forms of combination with one another (Latour 1999), Latour emphasises how this results in the production of new grids of intelligibility that, when circulated back to the field, make possible new forms of action on the varied entities or agents that make up the field in question. Applying this perspective to Spencer's work will help to clarify how his museum, scientific and administrative practices interacted with one another in ways that helped to shape a new, distinctively racial logic for the Aboriginal domain. Through the varied roles he played – anthropological fieldworker, photographer and filmmaker, collector, lecturer, curator, writer and scientific administrator – Spencer played a key role in what Latour calls the 'circulation of reference' from sites of collection to centres of calculation and back again in the form of grids for administrative action on the colonial social. By producing the race as such as the transactional reality or 'working surface on the social' through which the government of Aborigines was to proceed, these different aspects of Spencer's work functioned as the moving parts of a new machinery of administration that brought about a new socio-material ordering of the management of Aboriginal bodies in both space and time.

Racial essentialism and governing strategies

Spencer's photographic practice was quite varied, including humanistic portraits of individual Aborigines that suggested a degree of intimacy and mutual respect between photographer and subject, as well as photographs informed by the conventions of the noble savage (Batty, Allen and Morton 2005). I focus here, however, on the photographs that he selected to accompany the paper he presented to the British Association for the Advancement of Science at its eighty-fourth meeting in Melbourne (Spencer 1914). I have chosen this text for a number of reasons. First, in terms of the type of photographs selected, the elements it assembles are similar to those in Spencer and Gillen's *Native Tribes of Central Australia* (1899) on which it draws, as they are also of the principles informing Spencer's arrangement of Aboriginal materials at the National Museum of Victoria.[12] Second, going beyond his and Gillen's fieldwork in Central Australia, he draws on the photographs and publications of other anthropologists – A. W. Howitt, W. E. Roth, John Mathew and Daisy Bates – for other parts of Australia, to present an account that, by bringing together different tribal practices, produces a new reality that underlies and, ultimately, discounts apparent differences between them. That unity is grounded in a shared bloodline that accounts for the Aborigines' status as – allowing for possible exceptions – 'the most backward race extant' revealing 'the conditions under which the early ancestors of the present human races existed' (Spencer 1914: 33). Third, as a text that was presented to, and circulated widely within, the national and international scientific communities,[13] Spencer's address played an important role in circulating the conception of a common racially defined Aboriginality produced by abstraction from a variety of 'field' sites. It was precisely this racialisation of Aboriginality that was eventually relayed back onto the Aboriginal inhabitants of those sites, subjecting them to an administrative grid that subordinated their specific differences to a common racial logic.

There are three main types of photograph assembled in the text. First, there are photographs of various ceremonies, rituals and medical practices from different tribes – the Urbunna's Tritichinna ceremony of the snake totem and the Arunta ceremony of the witchetty grub totem, for example – presented, side-by-side, as evidence of a shared level of cultural development. Second, there are various photographs of churingas, objects of magic, weapons, tools, utensils and shields. Taken, again, from different tribes in different parts of Australia, these testify to a shared level of material culture, thus associating the Aborigine with a set of stone-age tools at the primitive level of the hunter-gatherer. Third, but as the opening set of photographs, providing a racialised context for the interpretation of the photographs of ceremonies and objects that follow, there is a series of head-and-shoulder anthropometric portraits of naked Aboriginal men, women and children from different tribes whose distribution in different parts of Australia is shown in an accompanying map. While noting considerable differences in bodily appearance (height, breadth of nose, slope of forehead) and in bodily practices (presence or absence of beards, practices of scarification), Spencer argues that the colour of all these men, women and children is uniform ('the native is dark brown chocolate, not black') and rooted in a shared bloodline (he rules out, in this text, the possibility of 'the mixture of two races differing in colour' [34]).[14]

It is important to note, as Morphy (1996a) does generally of Spencer's anthropological practice, that Spencer by no means minimises the variety of practices between different tribes. To the contrary, this is a constant refrain throughout his text: 'there are very considerable variations in regard to totemic customs and beliefs in different tribes' (Spencer 1914: 43); 'there are endless varieties of clubs and spears' (78); 'Clothing and ornament vary very much in different tribes' (79). However, the terms in which he explains these variations retrieves them to a racialised production of the Aborigine. This variation is accounted for as a result of the dispersal of the tribes from shared ancestors, who observed common customs and forms of social regulation, to different parts of the continent where those shared ancestral traditions were adapted to different climatic and physical conditions of existence. However, such variation registers no departure from the constraints of a common bloodline connected to a shared stage of evolution or, more accurately, of arrested development. Spencer accounts for this arrested development in terms of the combined lack of natural and racial competition: unlike Europeans whose wits had been sharpened by savage natural competitors (lions, tigers, etc.), the Aborigine has had nothing more fearsome to compete with than giant diprotodons and kangaroos 'who were quite as anxious to get away from him as he was to capture and eat them'; and he 'has never had to contend with any higher race' (Spencer 1914: 33). The only differentiation Spencer allows is a temporal/racial one separating the Australian Aborigine from the Tasmanian who, in spite of a larger skull size, Spencer describes as being at a lower stage of development and, therefore, as certain to have been exterminated by the Australian Aborigines had the two races come into contact.

This commitment to a necessary racial unity is evident in a later paper in which Spencer puzzles over linguistic evidence suggesting a range of blood and shade divisions among Aboriginal tribes. Spencer pours cold water on this thesis by reporting his own comparisons of standard colour tints with 'the actual colour of the skin of very many natives in various tribes' (Spencer 1921: 4). He tells us that he could find no variation of colour within any tribe except for, in some case, minor ones between men and women, which, however, he discounts as the probable effect of men's use of darker pigments in bodily decoration. More distinctively, Spencer seeks to disqualify the linguistic evidence of colour variation by applying the colour tests that had been developed by W. H. R. Rivers in the context of the late-nineteenth-century debates over the colour blue. The impetus for these debates derived from William Gladstone's contention that the philological evidence suggested that the ancient Greeks had not developed the capacity to distinguish blue within the colour spectrum. This led to the development of tests to assess whether contemporary 'primitives', as the survivors of the past in the present, also lacked this capacity (see Dias 2004: 90–108). Spencer draws on this tradition, in which the capacity to separate blue off from other colours served as a sensory index of civilisation, to discredit the theory that the range of variation of blood and shade registered in Aboriginal languages might be the legacy of real distinctions of blood and shade. Reporting the results of his own applications of Rivers' colour tests, he claims that Aborigines were only able to distinguish between black, white, red and yellow; that they had difficulties in distinguishing black, brown and grey from each other; and that they were – in their natural state[15] – unable to distinguish blue. Spencer's conclusion is that the variations of shade registered in language use must be disconnected from the supposition that they reflected the perception of real colour differences as supposing 'a fine colour sense that at least our Aboriginal does not possess' (Spencer 1921: 4) and, in accordance with evolutionary accounts of the development of colour perception, could not have possessed in the past either. In this way, an enduring racial continuity rooted in a common bloodline is ensured in which the only difference in colour that is allowed is that between children and adults. 'Every child at birth is copper-coloured', Spencer tells us in a formulation that restates his position in his 1914 address, 'but in the course of a few days the skin darkens and assumes the chocolate brown of the adult' (50; see also Spencer 1914: 36). While granting Aborigines the ability to distinguish variations of shade within this blood category, variation can neither range more widely nor be perceived because of an essentialist racialisation that chains Aboriginal blood and eye together in a tight double-bind.

To place this racial production of the Aborigine in a proper historical perspective, it is necessary to understand the role it played in deleting – or at least weakening the force of – the earlier transactional realities or 'working surfaces' that had organised the interfaces between coloniser and colonised. The key role that Spencer accorded the tribe as a major unit of analysis is important here, comprising, as Kevin Blackburn (2002) notes, a significant departure from earlier traditions in Australian anthropology. These were typically concerned to identify the existence of different nations at the supra-tribal level: not, though, nations as sovereign political entities but, in the words of A. W. Howitt, as agglomerations of tribal groups that 'represent a social aggregate, namely a community bound together, despite a diversity of class systems, by ceremonies of initiation, which, although they vary slightly in different localities, are yet substantially the same and common to all' (Howitt, cited in Blackburn 2002).[16] It was in accordance with this biblical conception of the nation as a cultural group of common descent that Howitt and other contemporaries sought, well into the 1880s and 1890s, to map the relations between different Aboriginal nations as ones between different cultures, which, Howitt argued, differed in their languages, practices and identities as much as the Scots from the Welsh.

The significance of the tribe as a unit of analysis that was favoured internationally in the late-nineteenth- and early-twentieth-century development of fieldwork, with, Blackburn notes, Spencer as its leading Australian exponent, was threefold. First, it formed part of a new hierarchical distinction between societies that were nations and those that were tribally organised that was mapped on to a racially organised distinction between the civilised and the uncivilised.[17] Second, in place of the wider self-identifications that Aborigines had often testified to in earlier studies, and that, in the context of the land claims that have followed in the wake of the Mabo and Wik judgements,[18] they have since reclaimed, it substituted the tribe as a unit of colonial ordering that served to nullify cultural differences operating at a meta-tribal level, denying them any political or administrative significance. Third, it substituted a racial conception of Aboriginality as the common factor that cohered the customs and practices of different tribes into a single whole, thereby producing the race as such, even if through the mediation of different tribes, as the surface that government was to act upon.

Just as significant, however, indeed perhaps more so, was the substitution of the imagery of the Aborigine as a naked primitive for earlier photographic depictions of Aborigines as already well on their way to becoming self-civilising subjects. Jane Lydon's study of the photographs that were taken at Coranderrk Aboriginal Station over the period from its establishment in the 1860s through to the early 1900s is highly revealing here, showing how earlier conceptions of the Aborigine as possessing a human potential for improvement persisted into the early twentieth century. Located to the northeast of Melbourne, and within easy day-trip reach from the city, Coranderrk bore the imprint of a range of civilisational discourses – from Christian missionary ones to liberal ideals of self-improvement – in its commitment to providing a site in which Aborigines might, under white tutelage, become self-civilising subjects. The Aborigines who lived there had their own complex reasons for complying with this civilisational regime while also negotiating and imbuing it with their own distinctive values. But these are less my concern here than the ways in which the inhabitants of Coranderrk were presented in over 40 years of being photographed – by professional Australian photographers, by visiting Europeans and by their managers. I am, though, more particularly interested in the ways in which Coranderrk's Aboriginal inhabitants collaborated in the production of images of themselves as active participants in a civilising process, in view of their realistic assessment of the role that their circulation played in organising the interfaces between white and black Australia.

I shall present Lydon's analysis of just three of the photographs she discusses. The first, *The Yarra Tribe Starting for the Acheron 1862*, is a mythical, biblical framing of the initial establishment of Coranderrk. It depicts a staged re-enactment of the departure of the Yarra people for Coranderrk, with a group of fully clothed Aborigines – men, women and children – walking across a clearing in front of a bush setting. It thus marks their departure from a condition of nomadism set against a scene of nature to a place of permanent settlement. It is also clear from the tools they carry that they are setting off for a life of agriculture (whose absence, under the influence of Lockeian ideas, had served as the primary justification for the dispossession of Aborigines under the doctrine of *terra nullius*). Both their tools and their Western clothes – making them indistinguishable from the station manager who accompanies them – separate the Aborigines from the scene of untamed nature that is the backdrop for their journey. The second picture, Charles Walter's portrait 'Timothy', is of a middle-aged Aboriginal man from Coranderrk. Timothy is shown fully clothed and Walter deploys the conventions of European portrait photography, including Timothy's holding an opened book in his right hand, as a sign of his acquisition of literacy as a key marker of a capacity for self-civilising. The only photograph Lydon presents of an unclothed Aborigine is the half-naked 'Australian de Coranderrk' taken by the French photographer Désiré Charnay and eventually ending up, fittingly enough, at the Musée de L'Homme. Charnay's purpose had been to produce an anthropometric portrait after the fashion recommended by Thomas Huxley (see Spencer 1992) – that is, naked, and arranged against standard measuring devices – which, as we have seen, informed the scientific aspects of Spencer's photographic practice. The significance of this photograph in relation to Coranderrk, however, is its rarity: the inhabitants of Coranderrk usually refused to be photographed other than fully clothed as they were only too aware of the association of nakedness with notions of savagery. Yet it was an exception that confirmed the rule. Lydon tells us that Charnay paid a high price for the photograph, recording that this particular subject who, quite apart from the fact that he had clearly kept his trousers on, charged five shillings for the privilege.

If we compare Spencer's photographs with those from Coranderrk we can see how Spencer contributed to the development of a nature-culture in which Aborigines were, so to speak, jumped backwards in developmental time. For these images either post-date or are contemporary with those from Coranderrk. As such, they both represented and were instrumental in effecting the transition from one nature-culture to another: from earlier liberal humanist and Christian conceptions in which Aborigines were viewed as a part of a collective humanity that was distinct from nature and therefore as capable of being civilised and, eventually, of civilising themselves, to one in which they were forever poised at the stage of an incomplete transition from nature into culture. The same was true of the arrangement of the Aboriginal collections at the National Museum of Victoria in depicting Aborigines as occupying the position of an 'evolutionary degree zero' representing the base level from which human evolution elsewhere had proceeded, but not in Australia. This nature-culture formed part of a revival of earlier innatist conceptions associated with polygenetic conceptions of race in which the apparent 'unimprovability' of the Aborigine had been accounted for in terms of an absolute racial difference. According to Kay Anderson and Colin Perrin, this was later translated into a matter of being 'too late' as, in the interpretations of Darwinians like Spencer, the problem was not that Aborigines absolutely lacked any capacity for improvement. It was rather that, having languished for so long on the cusp between nature and culture, there was not sufficient time for them to acquire the material, intellectual and cultural resources needed to compete effectively with a superior race before the inexorable laws of racial competition led to their extinction (Anderson 2007; Anderson and Perrin 2007).

In the early years of the twentieth century, then, depictions of Aborigines were inscribed within two different nature-cultures that jostled for space with one another within Melbourne's public visual culture. At a time when the National Museum of Victoria had been moved from the University of Melbourne to a more popular location in the city centre, so Coranderrk also attracted thousands of day-trip visitors annually from Melbourne, while Spencer's public lectures – involving lantern shows and film screenings depicting Aborigines as 'howling savages' – provided a popular evolutionary counterpoint to the Coranderrk experience. The disjunction between these two public versions of Aboriginality, it is important to add, cannot be accounted for in terms of a supposedly natural encounter with naked primitives when Spencer moved from Melbourne into the desert regions of Central Australia. As Wolfe notes, the readers of Spencer's and Gillen's texts would have had no sense that the Arunta had been brought under the influence of a Lutheran missionary ten years before Spencer and Gillen encountered them, or that they had earlier been photographed fully clothed alongside Strehlow and his family in texts that, since they were written in German, were little-known in either Australia or the English-speaking scientific community (Wolfe 1999: 155). Spencer reflected revealingly on these issues in a letter to a fellow anthropologist, the Reverend Lorimer Fison. Expressing his concern as to whether 'it was fit and proper to exhibit in public cinematograph pictures of natives who do not wear all the clothing that they ought to', and noting that he could show slides 'to which not even the most strict Presbyterian elder could take exception', he argues that this could only mean 'the elimination of the best of things. ... the real native' (Spencer, cited in Cantrill 1982: 37).[19]

Racial, individual and governmental times

I have drawn on Latour's account of the role of collecting institutions within the processes through which worlds are made and mobilised in view of the emphasis it places on assembling as a practice, which, by rendering the things it brings together readable, combinable and presentable with one another in new ways, makes new realities thinkable and actionable. As we saw in Chapter 2, Tania Li, in commenting on the virtues of assemblage theory, similarly stresses how, in contrast to terms like apparatus and *dispositif*, which focus on what are often presented as settled or even completed formations, assemblage 'flags agency, the hard work required to draw heterogeneous elements together, forge connections between them and sustain these connections in the face of tension' (Li 2007: 264). A further advantage of the concept, she argues, is that of bringing together the 'material content of assemblages (bodies, actions, passions) and enunciations (statements, plans, laws)' (Li 2007: 265) on the same plane, linking them not hierarchically in relations of causal dependence or in the order of the real and its representations but as elements that are reciprocally inserted within and presuppose constitutive relationships to one another. Insofar as they relate to bodies, however, assemblages, as Deleuze and Guattari put it, constitute 'a precise state of intermingling of bodies in a society, including all the attractions and repulsions sympathies and antipathies, alternations, amalgamations, penetrations, and expansions that affect bodies of all kinds in their relations to one another' (Deleuze and Guattari 1988: 90).

The history of the relations between white and black Australia has witnessed a plethora of the coming together of statements, plans and laws with bodies, actions and passions to effect varied, strenuously resisted, orderings of the relations between black and white bodies – and the various shades in between – within Australian social space. This has, for the main part, involved compulsory rearrangements of the relations between racialised bodies brought about by the removal of Aborigines from their lands, camps, kinship and tribal structures, particularly via the abduction of Aboriginal children, and their relocation in sequestered environments to be subjected to varied forms of management (Commonwealth Government 1997; Haebich 2000). Spencer's address to the British Association for the Advancement of Science was delivered in the same year (1914) that a new way of arranging bodies in social space was introduced with the opening of the first institution in the Northern Territory designed expressly for 'half-caste' Aborigines.[20] This constituted a significant stage in the development of a new governmental rationality in which anthropological expertise was harnessed to the management of milieus by fostering new forms of somatic co-mingling and separation, which plucked the half-caste away from the full-blood in accordance with a new civilising strategy whose underpinnings were those of the absolutely racialised Other that Spencer's texts and collections enunciated.

The category of the half-caste was not, of course, a new one; but it had not previously been accorded any significance in the management of Aboriginal populations. Bain Attwood notes that although the distinctions between full-bloods, half-castes, quadroons and so on had long been recognised, they had, in Victoria, been denied any legal or administrative significance as late as 1877. When, a few years later, these categories were used as the basis for ejecting half-castes from Coranderrk, this was on the basis that they were not owed the same compensation or 'duty of care' as the full-blood Aborigines whose beckoning destiny was extinction. And in the missionary domain, at Hermannsberg, half-castes were regarded as inferior to full-bloods through into the 1920s, their skin colour functioning as a sign of spiritual dilution. Spencer, too, in the preliminary report he prepared during the year – 1912 – he served as Special Commissioner and Chief Protector of Aborigines of the Northern Territory invested no special positive value in the half-caste.[21] Indeed, he argued that they posed a threat of deterioriation because the Aboriginal mother 'is of very low intellectual grade' (axiomatically) while the white fathers were (usually) from the 'coarser and more unrefined members of higher races', with the consequence that their offspring were unlikely to be of 'much greater intellectual calibre than the more intelligent natives' (Spencer, cited in McGregor 1997: 80–81). Nor did Spencer at this time support separating half-castes in special institutions. He was adamant that no half-caste should be left in native camps, recommending that if they lived near or worked in towns they should be placed in compounds, and that they should be placed in reserves if they lived in remote areas in a 'wild condition'. However, both compounds and reserves were places where all Aborigines, half-castes and full bloods, were to be placed in order to receive education and training to lift them to a higher stage of civilisation.

In a later report he prepared for the Department of Home and Territories in 1922, Spencer followed through the logic of his 1914 address in investing a positive racial value in the half-caste. To ensure that they could realise their potential of achieving a higher level of development than the 'pure-blood blacks' (Spencer, cited in McGregor 1997: 145) he recommended the establishment of a half-caste station equipped with dormitories, bathrooms, dining rooms, school, smithy, carpenter's shop and other facilities for industrial and technical training. It would, he now argued, 'be a fundamental mistake and most prejudicial to their welfare to place [half castes] on a station along with full-blooded aboriginal children', urging that *'the half castes must be encouraged to marry amongst themselves'* (Spencer, cited in McGregor 1997: 146). Viewed in the context of the legal restrictions on inter-racial sexual relations that were introduced in

1922 to stem the increase in the number of half-castes, this constituted a programme that aimed to breed out the colour of the existing 'rump' (there were an estimated 624 half-castes in the Northern Territory in 1922) of past miscegenation.

The modes of interaction between white and black that were thus performed across these new institutional arrangements can only be understood in terms of the relations between racial time, individual time and the new forms of governmental time associated with the development of the Aboriginal domain. As we have seen, Spencer effects a new racialisation of Aboriginality in the role he accords race, rooted rigorously in bloodline and skin colour, to account for the shared substratum of Aboriginality that gives Aboriginal culture its underlying unity in spite of varying customs and practices among different tribes. At the same time, however, a dual temporality is attributed to Aboriginal culture as a result of the contradictory way in which Spencer inserts it into evolutionary time. On the one hand, it is a culture that lacks an inner temporality having been more-or-less static for millennia as a consequence of its grounding in a race that, lacking any competition, has failed to evolve. On the other hand, once faced with such competition as a result of European occupation, the Aborigine faces annihilation. Aborigines thus enter into evolutionary time for only a split second, at the very moment they are destined to leave it by being driven to extinction. Yet Spencer also provides for a different temporality at the level of the individual. While the laws of racial competition destine 'full-blood' Aborigines to extinction, managed inter-marrying among half-castes provides a route through which they and their offspring might survive, but only on the condition that they cease to be Aboriginal in the sense that is, for Spencer, defining and constitutive: that is, racially.

It is, to go back to Wolfe, this racialised production of the Aborigine as an absolute and undifferentiated other that formed the new 'transactional reality' underpinning the development of a new governmental time of assimilation in the context of the early-twentieth-century development of the Aboriginal domain. For it opened up a space in which half-castes could be separated from 'full-bloods' and absorbed into white society through a programme that was simultaneously one of progressive whitening and progressive civilising that would eventually prepare them for citizenship. It was precisely this civilisational and epidermal advantage that early national governmental rationalities sought to promote. As Tom Clarke and Brian Gallagan (1995) show, these distinctions provided the crucial terms in which the relations between the Aboriginal domain and the distribution of citizenship rights and duties were managed in the context of the independent national polity that was developed in the first two decades following the 1901 Federation of Australia. And in a series of photographs that formed a part of this new assemblage, Spencer depicted the half-castes brought together in these stations much like Timothy at Coranderrk, fully clothed and accompanied by the trappings of European civilisation – except that these were now markers of their incipient whiteness.[22]

5 Collecting, instructing, governing
Fields, publics, milieus

In his biography of Marcel Mauss, Marcel Fournier (2006) draws attention to two texts in which Mauss lamented France's lack of adequately developed institutions for the pursuit of ethnographic research and the dissemination of its findings. In the first, Mauss (1913) compares French ethnography unfavourably with British, American, Dutch and German anthropology. He particularly regrets its lack of a developed fieldwork tradition and, as his explanation for this, a failure on the part of French ethnographers to connect their work to the tasks of colonial administration. If it were to close this gap, French ethnography needed 'first, field studies, second museums and archives, and third, education', the latter directed towards the training of ethnographers. These steps were necessary, he concluded, if France were to fulfil its responsibilities to its colonial subjects, whom he characterised as hitherto 'the human groups it was trying to govern without even knowing them' (Mauss, cited in Fournier 2006: 167). In the second text (Mauss 1920) it is the relationship between ethnography and the French public that exercises Mauss' attention. Regretting that there was still 'no museum of ethnography in France worthy of the name' and 'no laboratories dedicated specifically to the study of indigenous peoples', he also complains that the 'general public know nothing of our research' and urges the need for ethnographers to 'do publicity, since a science can become popular only though vulgarisation' (Mauss, cited in Fournier 2006: 215).

Mauss was one amongst many of the intellectuals whose political lobbying, public proselytising and organisational work across a range of scientific associations eventually led to the establishment of the Musée de l'Homme (MH). This was a process that played simultaneously in a number of registers. It was, first, an important site for what amounted to a significant reorganisation of the French scientific field. Mediating the relations between Durkheimian conceptions of sociology and ethnography on the one hand, and earlier ethnological traditions rooted in comparative anatomy and the natural sciences on the other, it played a key role in fusing these into a new synthesis for which ethnology became the preferred term.[1] It also comprised, second, a significant component in a long-term historical transformation of the relations between French ethnology and colonialism as the orientation of the latter shifted, in Alice Conklin's terms, from a stress on France's *mission civilisatrice* to a stress on the *mise en valeur* of the colonies. This was a shift that, as I noted in Chapter 2, displaced the earlier concern to civilise the colonised with measures designed to 'alter the social milieu in which individuals functioned – rather than to act upon the individuals themselves' (Conklin 1997: 8). The MH's role in organising fieldwork expeditions to varied colonial contexts – French West Africa and Indo-China, and Greenland (a Dutch colony) – was crucially important in this respect. Taken together, the fieldwork–museum–laboratory relations produced by these developments established new networks for the flow of texts, persons, things and technologies to and fro between metropolis and colony as parts of new programmes of colonial administration.

The MH also constituted a beacon for the mobilisation of ethnology as part of a distinctive public pedagogy allied to the politics of the Popular Front at a time of a significant increase in migration to France from its colonies. Its central city location in the Palais de Chaillot, close to libraries and institutions of instruction rather than to churches or temples, was, as Jean Jamin (1998) has noted, important in this respect. So were its connections to the institutions of broadcasting and, through its complicated relations to surrealism, to the world of art. It also developed distinctive connections to the worlds of sport, particularly boxing, and commercial entertainment as a means of publicising its work among the popular classes. It was in these respects a significant rallying point for social-democratic and socialist opinion at a time of heightened racial tensions. It played a significant role in marshalling the anti-fascist alliances of the Popular Front whose support proved politically important, particularly after the election of the Leon Blum government in 1936, in securing the funding for the museum's establishment. The MH also, finally, played a significant role in the historical reconfiguration of the relations between Paris, regional France and France's colonies with regard to their positions in the governmental rationalities of Greater France. However, it performed this role only symbiotically in its relations to what Fabrice Grognet (2010: 431) calls its 'siamese twin': that is, the Musée des Arts et Traditions Populaires (ATP). This sibling institution was incubated, alongside the MH, in the Musée d'Ethnographee du Trocadéro (MET) in a process of reciprocal differentiation that occupied the greater part of the 1930s. Redefining the concerns of national folklore studies in the light of the more scientific developments in the field of colonial ethnology, the eventual emergence of the ATP as a separate institution from the MH differentiated (albeit not entirely) the objects of ethnology represented by French rural popular classes and traditions from those constituted by the ritual practices of colonial indigenes.

It is against the backdrop of these concerns that I shall examine the development of the MH and, to a lesser extent, that of the ATP over the late 1920s and 1930s. The processes that this involved lend particular force to the concept of the relational museum proposed by Chris Gosden and Frances Larson (2007). In discussing this in Chapter 2 I noted its conception of museums as parts of extended networks of texts, things, instructions, technologies and transport infrastructures whose analysis requires that the roles of directors and curators be decentred, seen as merely parts of such networks rather than as sources of controlling visions. Invoking this conception of museums in relation to the MH and ATP might, though, seem somewhat paradoxical given that both museums were established by charismatic directors – Paul Rivet and Georges Henri Rivière respectively – who imbued them with their own unusually strong sense of social purpose and commitment (Laurière 2008: 413). Both were undoubtedly significant figures who operated adroitly across the relations – between metropolis and colony, ethnology and folk studies, ministries of education and of colonial administration, field and laboratory – that informed the development of the MH and ATP. It is, however, these relations and the more anonymous social and political forces driving their transformation that will occupy the centre of my attention. While these need to be tracked from the 1890s through to the 1930s to be placed in an adequate historical perspective, I shall zero in on the period from 1928 – when Rivet was appointed director of the MET followed, shortly thereafter, by the appointment of Rivière as his deputy – through to 1937, when the ATP assumed a life of its own,[2] shortly before the MH itself was officially opened in January 1938. In doing so, I note a second paradox. It is this decade that has attracted by far and away the most attention from museum and cultural historians, enjoying a paradigmatic status for the analysis of these two institutions in spite of their considerably longer histories.[3] It was, however, a decade in which neither of them yet existed. What existed, rather, were their programmes and the processes through which these were progressively shaped into being via the reformation – in terms of conception, function, design and layout – of the MET. It is, accordingly, the processes responsible for this reshaping that will be the hero of my tale here. Some of these were close to the MH: those concerning the reorganisation of the relations between scientific associations, the University of Paris and the arrangements for the administration of French museums that accompanied the establishment of the MH. Others concerned the changing governmental rationalities informing the relationships between France and its colonies, and Paris and provincial France, in the context of the governmental rationalities of Greater France. I begin with the latter.

Colonial humanism and Greater France: colonial and regional governmentalities

I noted, in the previous chapter, the influence that Spencer and Gillen's fieldwork had on Durkheim's account of the elementary forms of religious life. That influence extended also to Mauss (Durkheim's nephew). Both uncle and nephew had reviewed *Native Tribes of Central Australia* separately in the 1902 issue of *Année Sociologique*, and the book figures prominently in a joint review article on primitive classification they published in the following year. Mauss also had a detailed knowledge of the English school of anthropology from which Spencer took many of his intellectual bearings, and he developed personal connections with Tylor and Frazer during a visit to Oxford in 1898.

In spite of these indirect affiliations, however, the political and colonial contexts in which Australian and French anthropology developed in the early twentieth century were significantly different. The work of Spencer, as we have seen, was located at a moment of Australia's (relative) separation from the imperial domain of Great Britain through the establishment, after Federation, of a relatively autonomous national governmental domain. Australian anthropology was also shaped chiefly by its relations to the governmental rationalities of the internal forms of colonialism between a white, chiefly Anglo-Celtic, settler population and the Aboriginal population.[4] French colonialism, beginning in the 1880s, but especially after the 1914–18 war, was shaped by a different dynamic, particularly in Africa. It was, first, a colonialism based on overseas possessions, and one that, moving beyond the phase of wars of conquest, sought, in various ways, to make the colonies places to be populated (by the French) and whose indigenous populations were, in their turn, to become parts of an enlarged conception of France. Extending the conception of France beyond the limits of the Hexagon, or mainland France, the conception of Greater France sought to effect a union of peoples, cultures and territories by enfolding the populations of France's colonies into an extended conception of nationhood. At the same time, however, this enjoined the task of differentiating those populations. However much they were to be welcomed into the family of Greater France, the colonial populations were not – on the whole – regarded as suitable candidates for citizenship rights.[5] This was so, moreover, whether they remained in the colonies or whether, as they did in increasing numbers, particularly in the 1930s, they migrated to France. With regard to the former, the emphasis was increasingly placed on the *mise en valeur* of the colonised as a resource to be exploited for the furtherance of the economic and military prowess of Greater France. With regard to the latter, it meant differentiating immigrants from different parts of France's overseas empire in terms of their civic statuses and suitability for different kinds of employment (Blanchard and Deroo 2008).

How, then, to manage a new set of relations between an expanded conception of France to effect, simultaneously, a unity of territory and population – an extended people-nation, a Greater France – alongside a division between citizens and non-citizens: this was the governmental problem to which the MH responded and by which its practices were shaped. It was, however, by no means the only such response. To the contrary, it was merely one *dispositif* among many (the cinema, radio, travel literature) promoting a new culture of colonialism as a central aspect of French identity while simultaneously negotiating a subordinate position for the colonised as exceptions to the post-revolutionary traditions of French universalism.[6] The position of the MH among these *dispositifs*, however, was distinctive owing to the respects in which its management of the cultural flows from colony to metropolis, and its role in the administrative flows going in the opposite direction, placed it at the intersections of the processes through which the work of 'making culture' and 'changing society' was articulated across metropolitan and colonial contexts.

Patrick Wolfe's account of the general changes in anthropology's epistemological frames of reference that were brought about by the end of the
frontier or conquest phase of colonialism as, in varied ways, the colonised
were relocated within the governmental orbit of the colonial state is helpful
here. 'From the wholesale triumphalism of the expanding frontier', he argues,
'colonialism shifted to a diffident posture, offering indirect rule and fostering
local autonomy' (Wolfe 1999: 43). The virtue of this epistemological shift,
fully discernible only in the inter-war years, was its plasticity. By removing the
colonised from the frameworks of universal narratives of impending
improvement (or extinction), and by stressing their self-generating qualities as
synchronous totalities, colonial populations could be governed more circum-
stantially in ways that could be more easily adapted to different histories
and forms of colonial rule. The premium that this placed on the acquisition
of a detailed knowledge of the Other – dispersed now into territorially
differentiated cultures rather than distributed as stages along a continuum
of evolutionary time – led to a realignment of the relations between
museums, as centres of collection and calculation, and colonies, as sites of
collection (Boëtsch 2008). This realignment was principally brought
about by the role that was accorded fieldwork of regulating the flow of
expertise and instruments of collection from museum to colony; of texts,
objects and anatomical remains back from colony to museum; and of the
personnel and practices of colonial administration from the metropolis back
to the colony.

As we saw in the previous chapter, the fieldwork phase of anthropology
involved a new set of agents operating as parts of emerging scientific–
administrative assemblages that linked museums to colonial locations as sites
of collection that were also developing as governmental domains. But the
modes of its application differed significantly depending on the rationalities
brought to bear on those domains. While partly representing the differentiating focus that Wolfe attributes to the later phases in the development of the
fieldwork paradigm, Baldwin Spencer's racialised version of evolutionary
theory operated on the race as such, placing the Aborigine on the wrong side
of the biopolitical dividing line between 'make live' and 'let die'. The logic of
mise en valeur, by contrast, meant that, in the French case, the colonised were
regarded as a valuable resource whose labour was necessary to transform the
raw materials of colonial territories into economically exploitable forms. It
thus fell on the positive side of biopower, as a resource that was to be made to
live through varied forms of action (medical, sanitary) brought to bear on its
relations with its milieus at the same time that it was subjected to coercive
and punitive forms of labour control.[7]

Both Benoît de L'Estoile and Gary Wilder argue that the concern with
the distinguishing properties of indigenous cultures that typified the new
tendencies in French anthropological practice constituted a form of 'colonial
humanism'. They are both clear, however, that this is not to be understood
as simply a new humanistic ethic indexing the development of benevolent

forms of administration. They see it rather as, in Wilder's terms, an 'administrative-scientific complex' (Wilder 2005: 61) – one in which the MH played a pivotal role – which, drawing on the resources of both Durkheimian sociology and Maussian ethnography to develop an understanding of indigenous society as a distinct totality, aimed to secure economic development via the provision of social welfare while maintaining political order. This resulted in a number of instabilities and contradictions. Perhaps the most significant of these concerned the role of anthropology in the study of village life. This was identified as a set of customs and practices that needed to be known if African cultures and societies were to be preserved as the locus of traditional economic skills and organisational forms, and thus serve as a key resource for economic development. At the same time, village life was seen as a barrier to economic modernisation and thus as needing to be transformed. Wilder summarises the effects of this dual orientation as follows:

> The administration sought to transcribe native customs in order to allow colonial subjects to live within their own indigenous communities. Yet the very practices of ethnographic documentation worked to change those native customs in order to promote social transformations.
>
> (Wilder 2003: 237)

The ways in which such transformations were to be effected accentuated the differences between metropolis and colony. The development of social welfarism in inter-war France, Wilder argues, increased the significance of governmental practices that operated through the management of milieus relative to, but without displacing, those operating via the mechanism of the public to bring about transformations of individual behaviour. When translated to the colonies, however, the lack of civic institutions or of anything approaching a public sphere, and the absence of institutions of democratic governance meant that the necessary transformations of customary ways of life was to be effected through the scientific–administrative application of ethnology to the management of colonial milieus.

Both of these contradictions were subtended by the division at the heart of Greater France between citizens, a status reserved mainly for the French, and France's colonial populations. The reasons that were advanced to justify the denial of citizenship to the latter testified to the continuing influence of earlier evolutionary conceptions on the projects of 'colonial humanism'. Political and civic equality was denied the colonised not because their culture was inherently inferior but because – in a late echo of Tylor's notion of survivals – it had not yet entirely shed the legacy of its primitivism. As Wilder puts it:

> Like Mauss's ethnology, colonial humanism's antiracism depended on an implicit evolutionary logic in which cultural difference was seen as a

matter of time. African society was not defined as inferior; it was simply late. Natives had a capacity for citizenship but were presently too different, insufficiently evolved to exercise it.

(Wilder 2005: 125)

Similar contradictions applied to contemporary conceptions of the distant regions within the Hexagon albeit with important differences. How much weight should be accorded to these differences, and how they should be conceptualised, are matters of some controversy. Opinions differ particularly with regard to how the relations between the MH and the ATP, and the significance of their eventual separation, should be viewed in the light of the longer history of the affiliations between anthropology, museums and France's colonial and regional populations. Shortly after its establishment in 1879 the MET became a significant institutional locus for the mediation of the relations between these two different populations, their cultures and their places relative to the Third Republic's programmes of industrial modernisation and nation building. On the initiative of Armand Landrin, one of its curators, the MET, originally focused solely on non-Western societies, opened two new exhibition galleries in 1884 devoted, respectively, to the ethnography of Europe and France. This had the consequence, Jamin (1986) contends, that regional popular cultures within France came to be presented as belonging to the same time as the primitive: as survivals of outmoded ways of life destined to be eliminated by republican assimilationist projects committed, at that time, to converting both peasants and colonial subjects into Frenchmen (see Dias 2006: 176).

For Jamin, Rivet's closure of the MET's Salle de France shortly after he was appointed its director and the subsequent separation of the MH and the ATP signal the unclasping of this museological approximation of the rural popular to the primitive. Grognet, in developing Jamin's argument, interprets this separation as having both an epistemological and a governmental significance. Epistemologically, it separated the domain of the French rural popular from that of natural history and comparative anatomy, which, in his estimation as well as Conklin's, remained a significant force at the MH where a continuing commitment to the exhibition of anatomical and cranial remains alongside the exhibition of differentiated cultures provided a racialised set of underpinnings for the latter. The location of the two museums in separate wings of the Palais de Chaillot severed the connections that had earlier been implied between French rural populations and racially primitive ones. It nationalised and sociologised the former as parts of 'le peuple français' (Grognet 2010: 431) who, although they were to be transformed, were nonetheless to endure in their (transformed) specificity rather than to (eventually) disappear through programmes of colonial assimilation. This museological differentiation of these populations – of rural France from the global primitive – was, Grognet argues, further complicated by the parallel establishment of museums focused exclusively on France's overseas colonies.[8] These developments reflected a governmental concern to manage the populations of Greater France by differentiating the claims and entitlements of colonial subjects from those of Frenchmen at a time when significantly increased immigration to France coincided with rising unemployment and increased social tensions between the French and their (would-be) fellow citizens.

While by no means taking issue with all aspects of the Jamin-Grognet thesis, Daniel Sherman disputes the contention that the institutional separation of the MH and the ATP resulted in so clear a separation of either their epistemological or governmental orientations. He takes a different cut into these questions, focusing less on the continuing influence of comparative anatomy and natural history at the MH than on the significance of its conception of ethnology as a science modelled on the fieldwork disciplines that took issue with the antiquarian orientations and amateurism of French folklore studies. Viewed from this perspective, he argues, the separation of the two museums and their differential relations to comparative anatomy should not obscure the 'durability of the museums' constitutive ties' (Sherman 2004: 677) with regard to their common endeavour to establish ethnology as a science. Rivière and Marcel Maget, a curator and the head of research at the ATP, both paid lip service to folklore studies while drawing on Rivet's conception of ethnology to wage war on the folklore tradition from within in order to establish its organisations and procedures on a more scientific basis. They particularly stressed the means by which the MH had responded to Mauss' urging that ethnography should vulgarise itself as a means of public instruction while simultaneously piggy-backing on the increased standing that ethnology had acquired in the French scientific field through the new set of institutional affiliations it entered into that connected it to the university. Rivière and Maget also followed the example of the MH by including, alongside its exhibition galleries, a laboratory function within the ATP as a means of connecting it to the tasks of provincial government.[9] The ambition of the ATP in this respect, Sherman concludes, was not dissimilar from that of colonial administrators and was characterised by a similar tension: 'to bring advanced technology and social organisation to rural agriculture while preserving traditional crafts and promoting tourist development' (674).

There is, then, compared to the situation in Australia at the point of Federation, a more varied set of relations between museum practices, anthropological fieldwork, the public and the administration of colonial milieus that needs to be unravelled in the cases of the MH and ATP. This is mainly owing to the wider range of differentiations (between metropolis and province, metropolis and colony, between one colony and another) that had to be taken into account in managing the relations between different populations and cultures within the shifting frameworks of Greater France. To unravel these relations further, however, requires that we now look more closely at the processes bearing more immediately on the emergence of these two museums out of the MET.

Museum–field–public

I look first at the development of the MH in the light of the broader transformations in the relations between ethnography and physical anthropology
that took place in early-twentieth-century France.[10] The latter, represented
initially by the Société ethnologique (established in 1839) and later by Paul
Broca and the Société d'anthropologie de Paris (established in 1859), was the
dominant tendency throughout the nineteenth century. Committed to the
'scientific' study of human races and militantly secular, it was conducted
largely outside the University and was sustained chiefly through networks of
private scientific societies and associations, most of them eventually acquiring
state recognition. While a rival Société d'ethnographie de Paris was also
established in 1859, and also received state recognition, its opposition to the
biological reductionism of physical anthropology was largely ineffective,
partly because its credentials were largely spiritualist rather than resting on a
conception of ethnography as a distinct fieldwork discipline. The MET, whose
establishment had been prompted by the Exposition universelle held in Paris
in 1878, was committed to a project of salvage ethnography that would
document primitive cultures before they disappeared. It was, however, largely
ineffective: it was shackled by a budget that made it impossible for it to
conduct any fieldwork or to arrange any training in the discipline (Dias 1991).

However, the tide began to run in the other direction in the early twentieth
century. This partly reflected the influence of Durkheim's sociology, which
was translated by Marcel Mauss and Arnold van Gennep into the concerns of
a reformed Société d'anthropologie that was open to ethnography as well as
to physical anthropology; and it partly reflected the influence of a number of
new journals (the *Revue des études ethnographiques et sociologiques,*) and
societies (the Institut ethnographique international de Paris [1910], the Institut
francaise d'anthropologie [1911], a resurrected Société d'ethnographie [1913,
the earlier one had closed in 1903] and the Société des amis du musée
d'ethnographie du Trocadero). However, the key event was undoubtedly the
establishment of the Institut d'ethnologie as a part of the University of Paris
in 1925. Established with the support of the Ministry of Colonies and under
the joint direction of Mauss, Rivet and Lucien Lévy Bruhl, this was a key
moment in the reorganisation of the French intellectual field. It brought
together sociology, anthropology and philosophy in the persons of its
three leading *savants* and, through them, the institutional power of the
Sorbonne, the École Practique des Hautes Études and the Muséum national
d'Histoire naturelle, which was responsible for the administration of the MH
and where, in fact, Rivet's office as Director of the MH was located.[11] Its
primary purpose was to train a new generation of ethnographers and colonial
administrators who, schooled in the new humanism of Maussian ethnography
and deriving a knowledge of their subjects-to-be from the rigours of fieldwork, would nourish the development of a 'colonialism of the left'. To this
end, it was to bring its resources to bear against the excesses, or inadequacies,

of earlier civil, missionary and military colonial administrators while similarly acting as an intellectual and organisational alliance that would counter the influence of Louis Marin[12] and the École d'anthropologie, whose conservative conception of anthropology's role was a potent political force from the 1890s through to the 1930s. Rivet's activities in the political field in standing as a Popular Front candidate in the metropolitan elections were important in this regard. This helped in forging connections between the scientific and political fields that proved a significant counterweight to Marin's role as representative of the Front de la Liberté in the Chamber.[13]

These, then, were the changing institutional, political and intellectual coordinates that brought fieldwork into the centre of debates in the social and human sciences in France, and which made the MH a focal point for those debates. It was a nodal point in a network of institutions – 'a sort of scientific Grand Central' (Conklin 2008: 260) – that was to prove just as crucial in training the next generation of French anthropologists and colonial administrators in the Durkheim–Mauss tradition as it was to the task of instructing the public in its cultural and political implications. However, this did not entirely displace the position of physical anthropology, which, although no longer resting on the same anthropometric base that Broca had placed it on, remained central to the practices of the MET throughout the 1920s and 1930s. Indeed, Rivet's own practice (he had trained as a doctor and was schooled in Broca's anthropometric methods [Conklin 2002: 33]) retained a strong anatomical focus and, under his direction, the MET and, later, the MH continued the tradition of Broca's Laboratoire anthropologique in collecting and exhibiting a range of anatomical parts (see Hecht 2003).[14] In Alice Conklin's interpretation, the result was a constitutive tension between humanistic displays committed to celebrating 'the fundamental unity of humankind and the equal value of all cultures' and, side-by-side with this, the seemingly anomalous display of skulls and other anatomical remains reflecting the continuing influence of the assumption that cultural differences could be attributed to measurable differences in skull and body types. While this tension ran throughout the MH, the balance between these different orientations varied across its different galleries. Biological racialisation was most prominent in the introductory Anthropological Gallery where the visitor 'discovered the origins of humanity and the distinctive morphological, physiological and anatomical traits of modern humans and then proceeded to a display of the principal races through skeletons and skulls' (Conklin 2008: 263). These concerns were carried over, but as a subordinate theme, into a series of territorially defined galleries (the Black African Hall, the White African Gallery, the European Gallery, the Asia Hall, the Arctic and the Americas, and Oceania) via the display of racially differentiated skulls and skeletons alongside ethnographic collections arranged to display, as the primary message of these galleries, the distinctive qualities of different tribal or regional cultures. There was a variable balance between these elements within specific vitrines with, Conklin suggests, skulls being more likely to accompany arrangements of cultural artefacts in the Black African Hall, reflecting the conventional placing of African peoples on the lowest rungs of evolutionary hierarchies (Conklin 2008: 267–8).

In these respects the MH brought together materials that belonged to two different but overlapping disciplinary and institutional formations: those of physical anthropology and ethnography with, as we have already noted, the latter as the dominant element in the reinterpretation of ethnology that Mauss and Rivet promoted in the Institut d'ethnologie. These materials had, in turn, formed parts of different anthropological assemblages, the one placing a premium on osteological and craniological comparisons of skulls and skeletons for evidence of (as the case may be) underlying human unity and continuity, racial difference or evolutionary sequence, and the second on textual (in the form of photographic, film or sound recordings of ritual practices) and artefactual evidence of cultural differences. The ambition to overcome this division was particularly clear in a paper that Rivet co-authored with Paul Lester and Rivière in which, outlining the case for moving the Laboratoire d'anthropologie from the Muséum national d'Histoire naturelle to the MH, great stress was placed on the role this would play in realising the underlying unity of French anthropology by bringing together its osteological and ethnographic collections (Rivet, Lester and Rivière 1935).[15] This integrative function was, indeed, central to Rivet's conception of the MH as an assemblage of all of the materials that had previously been kept apart in separate collections. It should aim, he argued, to 'assemble together for a common task all the organisations, all the libraries, all the dispersed collections, which concern the races and human civilisations'.[16] He was just as concerned that it should aim to be an assemblage of different types of collection including a *bibliothèque*, a *photothèque* and a *phonothèque* alongside its material culture and anatomical collections.[17] This was central to its purpose of presenting both the unity of man – a unity underlying differentiated types – and the plurality of cultures and civilisations.

The importance assigned to these collections was closely related to their role as repositories for the materials gathered from colonial sites of collection. In contrast to the earlier history of the MET, the MH, in collaboration with the French ministries responsible for the administration of colonial affairs, organised a number of fieldwork missions in the 1930s, the most influential – albeit by no means typical – being the 1931–3 Dakar-Djibouti mission.[18] There were, however, profound differences between the missions that proceeded from the MH and the expeditions that Baldwin Spencer conducted from the National Museum of Victoria with regard to what I have called elsewhere their fieldwork *agencements*: that is, the range of agents (human and non-human) involved in fieldwork expeditions – the forms of recording technologies used, the gifts used to elicit the cooperation of the fieldworker's subjects, the fieldworker's tent, in lieu, famously, of the missionary's verandah[19] – and the different kinds of agential capacities these derived from their relations to one another (Bennett 2013). Some of these differences concern the different epistemological orientations of French anthropology from the Anglo-Australian formation represented by Spencer, while others have to do with the more developed set of connections between field, museum and public that informed the conception and organisation of the MH's fieldwork expeditions. I amplify these points more fully later but limit myself here to three aspects of the relations between the MH and its missions.

The first point to note, and it is a significant contrast with Spencer, is that the roles of museum director and anthropological fieldworker were not combined in the same person. The role of the director/curator – for this applied with equal force to Rivet at the MH and to Rivière at the ATP – was less to conduct fieldwork than to orchestrate it by coordinating the arrangements between varied agents. The French model aimed not to overcome the division between theoretician and fieldworker by combining these in the same person, but to improve the division of labour between these two roles, taking the nineteenth-century model beyond the artisan stage in which each *savant* had dealt with his own personal network of colonial correspondents to a more scientific and institutionalised division of labour (L'Estoile 2007: 103–16). When Rivet visited sites of collection, he did so primarily as an intellectual-administrator whose aim was to develop an infrastructure – by assisting the development of museums in Dakar and Hanoi, for example – that would produce a more efficient division of labour between museum and colony by establishing the MH as the coordinating centre of an institutional network of museums operating in different sites of collection (L'Estoile 2007: 118–30). The relations between Rivet and Mauss are also important here in shaping a conception of the MH as the headquarters from which expeditions would receive their instructions on what to collect. Mauss' earlier proposal for the establishment of a bureau d'ethnographie had a formative influence on Rivet in this regard (Conklin 2002; Sibeud 2007). The instructions for the expeditions focused on the need to collect, whether in object form or via recording devices, everyday and typical things and practices rather than the beautiful or curious. The Dakar–Djibouti mission, Rivet and Rivière (1933) pointedly reported, returned 3,500 objects, 6,000 photographic negatives, 200 sound recordings, notations of 30 separate languages or dialects, 300 Ethiopian manuscripts for the Bibliothèque Nationale, a collection of paintings and church murals, a zoological collection for the Muséum national d'Histoire naturelle, and 1,500 *fiches d'observation*. Replication of its accomplishments in these regards by other missions would result in an accumulating, but territorially and culturally differentiated, archive of man made available by the MH for scientific analysis.[20]

Unlike the earlier phase of armchair anthropology, however, this work was to be undertaken not by the individual *savant* but by a corps of experts in a specially segregated scientific setting. This led to a second significant difference: the break, particularly in the fieldwork practice of Marcel Griaule, the leader of the Dakar–Djibouti mission, with an individualised conception of fieldwork practice in favour of a division of labour between different specialists with a view to bringing a multi-perspectival orientation to the constitution and pursuit of their object of knowledge. In his methodological introduction to the 1933 issue of *Minotaure*, in which the findings of the Dakar–Djibouti mission were first put into broader public circulation, Griaule contrasts what he calls the extensive approach to fieldwork with the intensive immersive paradigm first introduced, he argued, by Frank Hamilton Cushing.[21] While by no means dismissing the immersive paradigm – although, as we shall see, this had a quite specific meaning and function for Griaule and many of his contemporaries – Griaule, invoking the urgency of the salvage paradigm, argued the need for a mode of study (and of collection) that would be 'sure and rapid' (Griaule 1933: 8). He accordingly dismissed the model of the ethnographer who does everything – 'l'ethnographe-á-tout-faire' – as out of date. If ethnographic phenomena were to be captured in their rounded entirety in order to understand a culture as whole, then – given the impossibility of the same person taking notes or making drawings while filming or taking photographs, or being able to observe ritual performances from different points of view (those of its male and female participants, for example) at the same time – the need for a coordinated division of labour among multiple specialists was paramount.

Perhaps the most distinctive aspect of the MH's missions, however, consisted in the ways in which these were planned, managed and executed – from their conception through the moment of their departure from Paris to their arrival at their destination(s) and subsequent return to Paris – with a view to the different kinds of contributions that their collections would make to both the MH's public and civic functions and its scientific and administrative ones. As I have focused mainly on the latter to this point, let me say a little more about the factors that shaped the former. These included significant changes in international museological practice and debate. The Office internationale des musées (OIM) played an important role here. Set up in Paris in 1926 under the auspices of the Society of Nations' Commission internationale de cooperation intellectuelle, it played a significant role in prompting European museums of art, ethnography and archaeology to transform themselves – on an American model – into instruments of popular and democratic education (Gorgus 2003: 72–82). In Rivet's case, this, allied with his strong commitment to the Popular Front, resulted in a conception of the MH as an instrument for reshaping French attitudes towards its colonial subjects by detaching the latter from evolutionary and hierarchical conceptions of difference and installing them in the space of a new humanistic universalism in which a common biological and anatomical substratum was overlaid by differentiated, but ostensibly equal, racial types and cultures. These developments formed part of a broader reshaping of the civic function of museums with all collecting institutions coming under increasing pressure to dedicate their exhibition functions to pedagogic programmes directed at the popular classes in ways that ran directly counter to the predominant tendency of nineteenth-century French museum practices (Poulot 2005: 143–6).

The ways in which texts and artefacts were assembled in new exhibition formats was a significant aspect of these processes. Rivière's role was crucial here in drawing together a number of contemporary exhibition practices – particularly those of the Scandinavian open-air museums, the development of the museum as an instrument of revolutionary instruction in communist Russia and American conceptions of museums as agents of popular democracy (Gorgus 2003: 21–31, 83–98) – to develop principles of display designed to give a clear, accessible and holistic picture of the interaction between the elements making up the whole way of life of different territorially defined cultures. Exhibitions in the MET had mainly been organised in accordance with decorative principles: as trophy collections in which the Other was depicted as a colonial possession, simultaneously exotic and primitive. Metal rather than wooden vitrines for greater transparency; the organisation of clear lines of sight for all objects; the provision of documentary and photographic information to relate objects to their regional milieus; the installation of vitrines in a modernist architectural space: in these ways, by contrast, the MH aimed to embody the principles of a *muséographie claire* that would exhibit ethnographic objects as the ordinary and typical indices of a culture in order that they might serve as the instruments of a new public pedagogy (Gorgus 2003: 56–60).[22]

This conception of the object's destination informed all aspects of the MH's fieldwork missions. These were organised explicitly with a view to the materials they collected being put on show for the edification of the French public. This intended destination was extensively rehearsed in the publicity through which the MH marked the departure of its missions; in the periodic newspaper reports that were based on the press releases summarising the reports that the ethnographers in the field sent back to the MH;[23] and in the publicity campaigns accompanying the return of the missions and the special exhibitions prepared on the basis of their collections.[24] Jean Jamin nicely captures the circuit that bound field and museum together in his discussion of the tournament featuring the American boxer Al Brown, at that time the bantamweight world champion, which the MH arranged in 1931 as both a publicity and fundraising event for the Dakar–Djibouti mission. Noting that uniformed museum guards were posted at each of the four corners of the boxing ring, Jamin argues that this exhibition of Brown under the surveillance of the museum anticipated the eventual destination of the 'objets nègres' the mission was to collect and return to the MH for exhibition under the watchful eye of its guards (Jamin 1982: 78). The point, however, is a more general one with two main aspects. The first concerns the respects in which the missions were significant pubic events, closely connected through the MH to the public sphere. The second concerns the high modernist ethos with which the MH's fieldwork missions were imbued. In reporting the MH's sponsorship of the tournament featuring the Al Brown match, *Paris-Midi* drew on Brown's association with America to tell its readers to think again if they still thought of the MET as an old, dust-ridden collection.[25] The MH's famous association with Josephine Baker operated similarly: the modern, as evoked by a leading representative of progressive American black culture, had arrived at the Trocadéro (Archer-Straw 2000). And it was as a representative of modernity that the MH set out to collect the cultures that it would later exhibit. The Dakar-Djibouti mission is again emblematic in this respect in the stress the public discourse of the MH placed on the up-to-date technologies of transport and collecting with which the mission was equipped.[26] Its specially designed lorries and custom-built demountable boat, its cameras, film and sound recording equipment: these were all 'of the moment' if not, indeed, stretching beyond it. Each mission would, on its return, be similarly linked to modernity through the regular series of radio broadcasts that Rivière established featuring talks by the leaders or members of the MH's missions: Marcel Griaule and Michel Leiris, for example.

This modernising ethos also informed the links between the MH's missions, its laboratories and their relations to its scientific–administrative objectives. It was through the connections that were thus established between field, museum and other components of the scientific–administrative assemblage of colonial humanism that the MH connected with the inhabitants of France's colonial territories not as publics to be transformed by educative measures but as populations whose behaviour was to be changed by conscripting scientific expertise to the task of engineering the milieus in which they lived. The establishment of the MH's 'laboratory' function was thus linked to the conception of the MH as an adjunct to the Institut d'ethnologie. As such, its purpose was to give the legislature 'a systematic knowledge of the customs, beliefs, and techniques of the populations it is called upon to direct' (Rivière and Rivet 1931).[27] This conception of its role is reproduced by representatives of all the key agencies involved: colonial and education ministries that provided its funding, the press, the curatorial staff of the MH and its field-workers.[28] The inclusion of a 'laboratory' function within the museum was not entirely new. Conklin (2008: 257–58) notes that it was a requirement that every institution operating under the jurisdiction of the Muséum national d'Histoire naturelle should have its own laboratory and research collection, and this had been true of the MET before Rivet's directorship. What was new, however, was how this laboratory function was expressed in the architectural distinction between the library and the *salles de travail* of different departments where selected collections were available to be worked with, all on the upper floors, and the location of the exhibition galleries on the ground floor.

Evoking the imagery of the laboratory in relation to these aspects of the MH's work was, in truth, more a matter of seeking a certain kind of scientific authority and validation rather than strictly emulating the procedures of the laboratory sciences. Such analogies, as we saw in Chapter 3, usually need to be treated with caution. Primarily textual, the materials that were brought together in this way – the photographs, films and *fiches d'observation* – mainly took the form of a set of files that could be assimilated to and coordinated with the files developed by France's colonial and overseas ministries. Through the research collections and the varied contexts that were supplied for working on these – by the Institut's students, academic and museum ethnographers/ethnologists, by the staffs of French ministries with colonial responsibilities, and by colonial administrators and military personnel – the MH formed a part of the developing complex of what Garry Wilder calls 'colonial ethnology' in which 'government policies were informed and produced by ethnographic knowledge just as ethnological science was informed and produced by administrative categories' (Wilder 2003: 221).

Governmental objects

There is, though, another aspect to the conception of the MH's laboratory function within the museum. This concerns its role in securing a status for the ethnographic object appropriate to the instructional tasks it was called on to perform in relation to the MH's role as an institution of popular cultural pedagogy. To address this, however, requires that we take account of a set of issues that, to this point, I have kept under wraps: those concerning the relations between anthropology and aesthetics within the ethos and practices of the MH, particularly those involved in its associations with surrealism. The terms of debate over these questions have, by and large, been set by James Clifford's account of the MH as the key institutional site for a distinctive intellectual fusion of ethnography and surrealism – 'ethnographic surrealism' – that shaped the direction of the cultural disciplines in 1930s Paris. The formations of 'ethnographic surrealism' were, Clifford argues, symptomatic of a more general 'modern cultural situation' consisting in 'a continuous play of the familiar and the strange, of which ethnography and surrealism are two elements' (Clifford 1988: 121). If ethnography, as represented by the 'ethnographic humanism' (135) of the MH, strove to make unfamiliar cultural worlds comprehensible, the aesthetic disposition of surrealism sought – in its combative relation to the taken-for-granted assumptions of Western culture – to bestrange the familiar.

These were not, Clifford stresses, mutually exclusive opposites. They rather defined the antinomies that structured and organised the crossovers between them as different practices combined elements of both, but in different permutations. Clifford singles out a number of such crossovers while arguing that the balance between 'ethnographic humanism' and the defamiliarising orientation of surrealism tilted in favour of the latter as, from the mid-1930s onward, the project of the MH began to be more clearly differentiated from that of the MET in which it was incubated. The inclusion of Michal Leiris, a leading literary light of French surrealism, in the Dakar–Djibouti mission alongside Michel Griaule; the sponsorship of that mission by arts organisations as well as by colonial ministries and scientific associations; the publication of the first reports from the mission on its return to Paris in the second issue of *Minotaure*, a surrealist publication established by André Breton and Pierre Mabille in 1933: these are among the various fusions of ethnography and surrealism that Clifford identifies. Rivière's involvements in Paris' jazz and *avant-garde* music scenes; the highly aestheticised principles of display he deployed in the exhibition of pre-Columbian culture at the MET before he was hired as Rivet's deputy; his participation in the contemporary enthusiasm for *l'art nègre*: these are also frequently cited to the same effect, as are the MH's connections, through Leiris, to the surrealist Collège de Sociologie (Arppe 2009).

These assessments have been called into question in more recent accounts. It is, however, less the empirical connections that Clifford traces between ethnography and surrealism than the methodological procedures underlying the interpretation he places on them that are at issue.[29] By interpreting the ethnography/surrealism conjunctions through the conceptual grid of his more general account of the art/culture system, Jamin argues, Clifford establishes a set of polyphonic connections between these different knowledge practices at the price of neglecting the specific conditions of their production. Drawing instead on a field analytic perspective, Jamin (1986) places greater emphasis on the differences between ethnography and surrealism, particularly with respect to their relations to the scientific and political fields. Ethnography, through the relations that were orchestrated around the Institut d'ethnologie and the MH, sought legitimacy in the scientific field as a form of university- and state-sanctioned knowledge of exactly the kind that surrealism parodied and rebutted. Equally, while both projects were anti-racial, they took up different positions in relation to colonialism: implacable opposition to all its forms and a demand for its immediate end in the case of surrealism;[30] a reformed colonialism, with the Institut and the MH as the key agent of its transformation, in the case of ethnography.

Jamin does not, though, deny the pertinence of a more limited set of connections between ethnography and surrealism. These took two forms: an initial set of contingent alignments, dependent on short-term personal contacts rather than on enduring institutional ties, through which the MH acquired a certain standing with Paris' artistic *avant-garde*; and a more lasting set of affiliations arising from the similarities between the French traditions of fieldwork that were developed in the inter-war period and the earlier, nineteenth-century voyage literature. The latter was a hybrid genre, installed ambiguously between the field of *belles lettres* and the developing field of ethnography, which narrated the journey into another culture as a form of self-discovery and transformation arising from the encounter with the Other. If the fieldwork tradition constituted a break with this tradition, it was an incomplete break to the extent that the missions authorised by the MH gave rise to narratives in which the ethnographer deployed this literary device to account for the kinds of transformative practices of the self they were led to perform as a consequence of their prolonged experience in the field. Here, the immersive paradigm of ethnographic fieldwork served less as a means for acquiring an objective knowledge of the Other than as an occasion for undoing and re-forming the self prompted by an experience of alterity. As such, it gave rise to a melancholic questioning of the values of Western civilisation and of the colonial rationalities that constitute the social and political underpinnings of the ethnographer's practice.

Vincent Debaene develops a similar argument in his account of the 'two books of ethnography' – that is, the formal scientific reports of the fieldwork missions in which the culture of the population under investigation is reported in a documentary fashion and a later, usually more extended text, which aims for a more literary and existentially fuller evocation of the 'atmosphere' of the culture in question. Michel Leiris's *L'Afrique fantôme* (Leiris 1996 [1934]), Griaule's *Les flambeurs d'hommes* (Griaule 1991 [1934]) and Alfred Métraux's *L'île de Pâques* (Métraux 1941) as accounts, respectively, of the Dakar–Djibouti mission, Griaule's earlier mission to Ethiopia and Métraux's mission to Easter Island are among the chief cases he has in mind. As Debaune puts it:

> The ethnologists were constantly caught between two conceptions of their work: on the one hand, in the name of objectivity and as a counter to the picturesque, they demanded that their labours be conducted in a strictly documentary manner, never forgetting to refer them back to a monographic inventory and to the museum collections; on the other, they never ceased to deplore the insufficiency of the document and its incapacity to reconstruct the 'atmosphere' of the society under investigation.
>
> (Debaene 2010: 121)

It is, however, Debaene's insistence on a clear separation between this aspect of the ethnographer's practice and the inscription of the texts and artefacts collected by such expeditions within the institutional practices of the MH that is of most concern to me here. Here, he argues, the imperative to establish the scientific credentials of ethnography entailed as clear as possible a separation of the manner in which such collections were processed and institutionally mobilised from the *belle lettrist* associations of voyage literature. The key problem this came to revolve around consisted in securing the epistemological status of the ethnographic object as a scientifically validated document of the culture to which it referred. This was achieved by the fusion of 'the thing itself (taken from its place and presented without alteration) and the lesson of things (explanatory labels accompanying the exhibits for instructing the public)' (52) that the MH's practice produced. The educative task of the museum required that the ethnographic object should be disconnected from the 'voyage narrative which isolated spectacular facts from their context' in order to present them instead as parts of 'coordinated ensembles, reinscribing curious practices in a coherent system' (55). This required that the object's progress from field to exhibit be carefully managed along each step of the way. This was the role of the MH's 1931 *Instructions sommaires pour les collecteurs d'objets ethnographiques*, which emphasised the value of collecting ordinary and the typical objects that would serve as documentary indices of their uses in everyday life rather than exotic or extraordinary ones; of the guidelines for photographic practice designed to secure the documentary status of the evidence of ritual practices collected in this way;[31] of the laboratory work of scientific validation and classification, which secured the documentary status of the MH's collections, thus rendering them interpretable as evidence of the distinctive totalities constituted by a regionally specific culture and society; and of the exhibition galleries where the work of labelling and the layout of exhibits in accordance with the principles of the *muséographie claire* would guide the visitor to a correct relativising and humanist understanding of the cultures on display.

These relations between field, laboratory and exhibition gallery worked to secure a particular set of capacities for the MH's ethnographic objects as objects of knowledge of a distinctive kind. Their scientific validation distinguished them both from curiosities and from aesthetic collections of fine-arts objects by investing them with the distinctive epistemological value of the document.[32] Their value as document simultaneously imbued them with a distinctive moral force and authority as governmental objects of a particular kind. This force, as Christine Laurière interprets it, derived from the evidence of 'added value' that was coded into the ethnographic object. This 'added value', especially in objects that were the outcomes of particular techniques of production – of ceramics, metallurgy, arts and crafts, for example – consisted less in their documentation of the particular use values accruing to objects within particular cultures than in a universally shared capacity to creatively transform the material world through the coordinated deployment of intellectual, manual, technical and artistic skills. It was (as she calls it) this 'environmentalist conception of the object' (Laurière 2008: 416) in which the object bears the imprint of the socio-cultural environment that shapes it while also testifying to the capacity of human practice to reshape such environments that conferred on the MH's ethnographic objects a certain degree of strategic plasticity. While such objects were able to serve as a source of knowledge for the administration of colonial milieus, they could also serve as useful platforms for the MH's pedagogic engagement with the popular classes given the positive value placed on technical forms of creativity in artisanal cultures.

****

Although shorn of its Popular Front associations – particularly during the 1939–45 war when the MH was pressed into a different kind of state service during the German occupation[33] – the documentary status of the ethnographic object that was fashioned during the inter-war years was to have a long historical reach. But it enjoyed perhaps its greatest public prominence during the moment of its demise: the rearguard action mounted by its curatorial staff and leading French anthropologists in protest at what was, in effect, the dismantling of the MH and its pedagogical project as its collections were dissolved, packed up and shipped across the Seine to provide the Musée Quai Branly with many of its key marker objects.[34] This too was a state project, initiated by Jacques Chirac in the final years of his presidency, committed to a public pedagogy that would confirm and celebrate the equal value of all cultures. It was, however, a state project conducted under the sign of the aesthetic as the objects the Musée Quai Branly acquired from the MH and other ethnographic collections were selected and curated to stress their unique and exceptional qualities – their beauty – as testimony to a universally shared capacity for a more exceptional form of artistic creativity. They were, as a consequence, largely shorn of their documentary status as the typical indices of specific cultures. In moving from the MH to Quai Branly, the ethnographic object was reshaped to serve a different set of purposes as it passed from the jurisdiction of one form of cultural expertise (the ethnographic) to another (the aesthetic).[35] I cannot follow here the transformations in the governmental status of the ethnographic object that this entailed. The simple fact of their occurrence, however, provides a convenient transition to my concerns in the next two chapters where I turn to the distinctive forms of cultural expertise that have been associated with the history of the aesthetic and the governmental projects of 'making culture' and 'changing society' that these have authorised.

Part 3

Governing through freedom: aesthetics and liberal governance

Inter-text 2

I have, in Part 2, sought to illustrate the general principles of my argument through two case studies. These have exemplified how the deployment of specific forms of anthropological expertise across the relations between museums, fieldwork sites and public spheres have 'formatted' the social for different kinds of governmental intervention in relation to metropolitan and colonial populations. The tack I adopt in Part 3 departs from this case-study approach. My concerns are rather to pursue a more general set of arguments concerning the distinctive properties of aesthetics as a form of knowledge that has played a crucial role in orchestrating the relations between practices of government and those of freedom. My purpose is, first, to trace some of the discursive transformations through which the aesthetic acquired the capacities that have allowed it to function as such a pliable resource across varied practices of governing through freedom. These are the questions I address in the next chapter. I then, in Chapter 7, bring this perspective to bear on the associations between aesthetics and critique that I touched on in Chapter 1. I do so with a view to identifying the respects in which these depend on a specific, and paradoxical, kind of authority derived from post-Kantian aesthetics.

6 The uses of uselessness
Aesthetics, freedom, government

'For a notion which is supposed to signify a kind of functionlessness', Terry Eagleton writes of the aesthetic, 'few ideas can have served so many disparate functions' (Eagleton 1990: 3). He attributes this to the versatility that the concept has derived from the indeterminacy of its definition as a term that has always been installed between a set of related oppositions: between spontaneity and necessity, and particularity and universality, for example. Its specific meanings and uses have thus depended on how, in particular circumstances, it has been called on to mediate between such terms and the social values attached to them. In general, however, Eagleton sees these uses taking two main forms. In the first, by fashioning the bourgeois subject as, like the aesthetic construction of the work of art, autonomous and self-determining, it has nurtured the development of those forms of internal self-policing that bring about the bourgeois' self-directed compliance with the dictates of state and market. It has thus provided 'the middle class with just the ideological model of subjectivity it requires for its material operations' (Eagleton 1990: 9). At the same time, the aesthetic has generated a conception of 'the self-determining nature of human powers and capacities' that, in the Marxist tradition, bequeathed 'a vision of human energies as radical ends in themselves which is the implacable enemy of all dominative or instrumentalist thought' (9). Eagleton thus aims both to critique the aesthetic and to preserve it: to account for it as a distinctive historical discursive and institutional formation in which the uselessness that the aesthetic attributes to the work of art is invoked to serve particular forms of bourgeois utility; and to retrieve that uselessness for a politics that will critique all forms of instrumentalism in the name of a self-referential humanity.

Peter Osborne (2006) has similarly invoked the aesthetic in a more conjunctural critique of the instrumentalism represented by the New Labour cultural policies of the late 1990s and early 2000s. Chastising these for sacrificing the emancipatory potential of the arts by pressing them into service as vehicles for achieving access and equity or cultural diversity targets, Osborne also directs these criticisms against the 'cultural policy push' in cultural studies. The pragmatism this represents, he argues, received its theoretical consecration in George Yúdice's conception of culture as a resource that has now been governmentalised for a variety of purposes (Yúdice 2003).[1] Castigating this conception as 'reductive and theoretically reactionary' (Osborne 2006: 44), Osborne appeals instead to Adorno's account of the relations between culture and administration in which the uselessness of the aesthetic, no matter how residually, retains a negative critical force in relation to the actual forms of its administrative use (Adorno 1991). Like Eagleton, Osborne, who is fully aware of the actual history of its uses (Osborne 2000), invokes the uselessness of the aesthetic in the spirit of critique.

Both Eagleton and Osborne thus ultimately set up their stalls as 'authorities of freedom' whose 'pitch' is to guide certain practices of freedom, those of critique, by means of a secondary discourse on art, music and literature. I shall not follow them in this. My concern in this chapter is rather to examine how particular kinds of freedom have been made up, organised, distributed and consumed within the space produced by the aesthetic as crucial aspects of – and, in many respects, the paradigm for – liberal forms of government. My concerns in this regard connect with my discussion, in Chapter 2, on the diagram of the aesthetic. However I shall be less concerned than I was there with the socio-material entanglements of the aesthetic. I shall rather focus on the properties of aesthetic discourse with a view to showing how, in its Kantian and post-Kantian versions, the form such discourse gives to the concept of art's autonomy has helped to shape the patterns of its subsequent worldly careers.

I shall do so by interpreting philosophical aesthetics and the distinctive form of knowledge that it lays a claim to (briefly, to offer a knowledge of that faculty, the aesthetic, which is not itself a knowledge) as a form of expertise for adjudicating and directing those practices of freedom associated with judgements of taste.[2] This will provide the basis for my discussion, in Chapter 7, of the ways in which the entitlement to exercise the capacity for guiding freedom that this engenders is a point at issue in current aesthetic debates.

Addressing these questions will involve a certain amount of toing and froing between Kant's aesthetic and the earlier traditions – notably those of civic humanism and the state aesthetics of Christian Wolff – which Kant both drew on and transformed, and the subsequent uses into which the aesthetic has been pressed precisely as a result of the forms of autonomy that Kant secured for it. As a good deal of ink has already been spilt on these topics, I shall chart a selective route through them by focusing on four aspects of Kant's aesthetic: first, his conception of beauty as that which pleases without a concept (the beautiful as 'what is presented without concepts as the object of a *universal* liking' [Kant 1987: 53]); second, his conception of the disinterestedness of aesthetic judgement (the 'liking that determines a judgement of taste is devoid of all interest' [45]); third, his conception of the aesthetic as a reflective relation of the subject to itself without regard to any qualities of the object ('what matters is what I do with this presentation within myself' [46]); and fourth, his account of the role of genius in distinguishing aesthetic art from the mechanical arts as a form 'whose standard is the reflective power of judgement, rather than sensation proper' (173). I shall consider these in their bearing on Kant's account of culture as the aptitude that produces 'in a rational being an aptitude for purposes generally' in a way 'that leaves that being free', and which does so by freeing the will from the despotism of desires so that it might be guided by the purposes that reason chooses (319).

I look first at Kant's critique of the relations between Christian Wolff's aesthetics and the *polizeiwissenschaften* of the Prussian state. I then look at how this critique drew on, and transformed, the connections between the uselessness that had been ascribed to aesthetic judgement in the earlier culture of civic humanism. This paves the way for a consideration of the form of canonicity that the concept of genius bestows on the artwork, and the consequences of this for the production and consumption of art. I conclude by considering how Kant's autonomisation of art and of aesthetic judgement provided the preconditions for the subsequent history of its instrumentalisation. I both draw on, while also registering disagreement with, the work of Jacques Rancière for this purpose.

From police to liberal government

Wolff's *The Real Happiness of a People under a Philosophical King* economically summarises the relationship that ought to obtain between philosophy and government. The ideal is '*that if Kings or Rulers are Philosophers, or Philosophers Kings, then it is that the End of Society is obtained*' (Wolff 1750: 5) where that end is envisaged as an ordered set of social and political relations governed by the principles of police. The '*common Good*' is secured when the philosopher-king is able to bring about '*the highest Good, which every individual can attain to in this World, according to the different State he is in*' (Wolff 1750: 4). The role of philosophy was to apply the logic of subsumptive judgement in securing this end, bringing every particular case under the heading of a determinate concept in order that the requirements of the common good could be accurately and consistently determined from one case to the next. This required a rare degree of philosophical penetration:

> For the *discursive* Judgment must determine about what ought to be done in every Emergency, which is evidently not established by a bare Attention to the present Case; but gathered from it by Virtue of a Ratiocination: For no one can persuade himself that the very Idea of the particular Case contains what the End of Government, civil Happiness, directs to be done.
>
> There is a Degree of Penetration requisite, which in vain is to be looked for in one who is no Philosopher, for forming *distinct Notions* by Means of Reflection, for distinguishing Circumstances, which are mutually determinable, or which may only co-exist, and for *determining* properly in every emergent Case. And after having duly *determined* the Case, some certain Principle is necessary, by Virtue of which we may

come to learn, what ought to be done in that Case, that we may not seem to act contrary to the publick Safety, Security and Peace. ... That you may therefore with a Certainty of Judgement attain to what ought to be done in any given Case, *Philosophy* is necessary.

(Wolff 1750: 27–31)

The difficulty that judgements of beauty posed for Wolff concerned their implications for the relations between the lower sensate faculty of intuition and the higher faculty of reason. In the context of the Prussian police state of Frederick the Great, these faculties functioned as a coded reference to the relations between rulers and ruled: that is, between the enlightened absolutism of the monarch and the state bureaucracy on the one hand, and their unen-lightened subjects on the other. If beauty, as a form of perfection, can be sensibly intuited, what is the relationship between this and the rational knowledge of perfection achieved by the higher faculty of reason? Or, when translated into questions of governance: what is the relationship between the confused perception of the people if, in matters pertaining to beauty, they are allowed the power of judgement and the enlightened rationality of kings and philosophers in determining and legislating for the common good? Wolff's solution is to allow the existence of the intuitive or sensible perception of beauty only on the condition that it is subject to correction and revision by being brought under the influence of a rational appreciation of beauty gov-erned by definite rules of judgement. The higher faculty is thus to raise the lower faculty from its confused and dim appreciation of perfection just as enlightened rulers are to lift the unenlightened classes out of confusion and darkness through a didactic programme, legislated from above, that does not involve any self-activity on the part of the governed.

In this revival of the Hobbesian view in which the unification of the mani-fold results from the political act of sovereignty, the social order is governed from above – just as the lower faculty is ordered by the higher one – by the king and philosopher bureaucrats who administer the affairs of state in the light of their rational understanding of the common welfare. Government thus bypasses the individual members of civil society who are, so to speak, to be directed as to how to judge correctly rather than possessing this as a natural attribute or achieving it through their own activity. The role of the citizen is similarly to be led into willing obedience by learning, through public dis-cussion, to understand the reason that is embodied in the law. The common man, however, is to be led into blind obedience through a perpetuation of the politics of spectacle associated with the principles of sovereignty:

The common man, who depends on his senses and can barely use his reason, is unable to grasp what royal majesty is; but through the things which he takes in through his eyes and which affect his other senses, he knows majesty, power and force with an indistinct but clear concept.

(Wolff, cited in Caygill 1989: 139)

It was as a consequence of this denial of any generally distributed capacity for independent judgement that questions of aesthetics became so politically loaded in mid- and late-eighteenth-century Germany. Howard Caygill (1989) singles out two figures who helped to dismantle Wolff's philosophical apparatus by probing his account of judgements of beauty: Alexander Gottfried Baumgarten and Johann Gottfried Herder. Baumgarten's early work is conventionally regarded as remaining within the order of Wolff's system in seeking to establish a scientific basis for recognising the aesthetic as a distinct and separate faculty, thereby effecting a procedural subordination of the sensible to reason. In his later *Aesthetica*, however, the aesthetic operates as a field of independent judgement that is not automatically subservient to the higher court of reason: 'My recognition of the perfection and imperfection of things is judgement. Therefore I have the power of judgement' (Baumgarten, cited in Caygill 1989: 164). This opens up the aesthetic as an area of self-activity, albeit one that keeps judgement in a tutelary relation to reason, thereby also inscribing the people in the same relation to the state. Herder's *Sculpture* is more radical in locating judgement within a dynamic economy of the senses in which the privileging of sight associated with painting is discrowned by being subordinated to touch. This makes touch the organising centre for an economy of the senses in which society, rather than being 'ordered as if on a visual surface by a superior eye' (Herder 2002: 185), is produced through the acts of citizens as they form themselves into an object for their own contemplation through their relations to sculpture in what Jason Greiger, in his introduction to Herder's text, describes as a combination of 'spatial seeing and an anamnesis of touch' (16). Judgement, here, is a formative activity, breaking with the visual paradigm and its corollary, to cite Greiger again, 'of a passive, policed subject' (16).

These, then, are some of the relevant coordinates against which the *Critique of Judgement* needs to be set. Published in 1790, it played a key role in the debates leading to the 1806 Prussian reforms that marked the transition from the police to the legal state in view of the place it produced for the aesthetic by inscribing it in a conception of culture as process of free – but guided – self-making that is enacted in civil society. Coming after his critiques of pure and practical reason, the *Critique* completes Kant's alternative to Wolff's state-centred account of the subsumptive logic of reason. Whereas this requires that all objects are subsumed under universals, Kant's system is a relational one. The relations of attraction and negation between objects are generated by the systems of relations that are established by the dispositions of the subject arising from the exercise of the faculties of knowledge, desire and of feeling respectively. In order to reconcile his account of the faculties with the practice of freedom, Gilles Deleuze argues, Kant sought, in the case of each faculty, to identify a higher form of that faculty so that it might find, in that higher form, the law of its own exercise within itself and thus not be obliged to defer to any external authority. It is through this higher form that the subject is placed in charge of the world and the interest of reason is secured. For when, in the case of knowledge, 'the faculty of knowledge finds its own law in itself, it legislates in this way over the objects of knowledge' (Deleuze 1984: 5).

In the case of the faculty of feeling, however, the higher form of this faculty – aesthetic judgement, the sense of the beautiful – neither legislates over objects nor secures an interest in reason. It cannot do the latter since it is, by definition, disinterested and so is independent of both the speculative and practical interests that motivate, respectively, the faculties of knowledge and desire. It is not legislative since it is, again by definition, always particular and so cannot subject a field of objects to its exercise. The aesthetic judgement takes the form of 'this rose is beautiful' not 'roses in general are beautiful' (Kant 1987: 59) since such generalisations imply logical comparisons that are the province of the faculty of knowledge. Powerless to legislate over objects, 'judgement can only be heautonomous. That is, it legislates over itself' (Deleuze 1984: 48). But in doing so it provides a basis for harmonising the exercise of all the faculties. The supposition that aesthetic judgement is universal, and so communicable to all without the intervention of concepts, provides the basis for a common sense based on the free interplay of the faculties in which imagination and the understanding are brought into a free, undetermined accord with one another. Aesthetic judgement, as the higher form of the faculty of feeling, thus provides for the free subjective harmony between the faculties that is necessary if the other faculties are to perform the legislative role that Kant assigns them.

As is well known, for Kant, the higher faculty of feeling is split into two forms: the sense of beauty that is experienced in relation to works of art, and the sense of the sublime that is experienced in relation to nature. Deleuze, glossing this division and Kant's higher estimation of the sublime, notes Kant's dictum that 'he who leaves a museum to turn towards the beauties of nature deserves respect' (56). However, at this point, with the liberal subject of aesthetic judgement now almost fully assembled and, when translated to the English context, ready to set off either to the Lake District or to the nearest art gallery, I want to backtrack to consider the earlier history of the relations between the notions of uselessness, disinterestedness, aesthetic judgement and freedom associated with the practices of civic humanism. I do so in order to understand how Kant deployed these to undermine Wolff's system and how, in the process, he transformed them by incorporating them within a new economy of the subject.

The distribution of judgement and the limits of liberal political community

The general context for the eighteenth-century concern with the relations between taste and social ordering is provided by the need, post-1688, to reconfigure the relations between state and civil society. There are a number of considerations to be factored in here: the (relatively) unfettered growth of market society and, as a part of this, the undermining of feudal status and power hierarchies and the declining influence of sumptuary codes as a way of prescribing and publicly marking social distinctions; the continuing potential, in the aftermath of the Civil War, for sectarian religious divisions to produce turbulent social division; and the diminished power of sovereignty given the reduced authority of the monarchy and the continued potency of territorial divisions within the nation in spite (or because) of the enforcement of political unity on England, Scotland and Wales. If these are the factors that Paul Guyer (2005: 57) singles out for attention, Mary Poovey adds the collapse, by 1688, of William Petty's project of political arithmetic. As a governmental project that had aimed to substitute 'government through information' for the uni- fication of the manifold though the principle of sovereignty, this had proposed a rational classification of population by means of economic function as a grid for social ordering that was also intended to displace the social disordering of religious factionalism (Poovey 1994, 1994a and 1998).[3] Its failure, Poovey argues, meant that, unlike their continental counterparts, social theorists in Britain were unable to look to the bureaucratic procedures of *polizeiwissenschaften* to provide a mechanism of social ordering and looked, instead, to taste as a mechanism that might emerge out of the free activities of individuals in their capacity as consumers.

The variety of positions that were staked out in the context of eighteenth- century debates over taste is too wide to be fully reviewed here. Instead, I shall focus on two figures – Anthony Ashley Cooper, the Third Earl Shaftesbury, and Adam Smith – to trace the formation of the three aspects of eighteenth- century discourses of the aesthetic that bear most significantly on its later assemblage into technologies of liberal governance. The first concerns how the notion of the 'je ne sais quois' – the notion that the judgement of art is an ineffable, indescribable affair that cannot be codified in rules – is refashioned. Originating as a gesture of aristocratic dissent from the official codifications of judgement of the French Royal Academy, this comes to function as the aspect of judgement that topples the subsumptive rationality of Wolff's system. It also constitutes the aspect of the new definition of the work of art as unfathomable that accompanies art's transition from decorative artefact to a prop and resource for governance within post-Romantic aesthetic technologies of the self. My second concern focuses on how the notion of disinterestedness is reworked from its association, within the culture of civic humanism, with the independence of the landed gentry into the 'purposiveness without purpose' of the Kantian conception of aesthetic appreciation and the role this plays in Kant's conception of self-governance as a practice of freedom. I look finally at the influence of Shaftesbury's conception of the role of aesthetic judgement in opening up a new space within the self – a space of self-inspection and self- reform – whose uneven social distribution becomes one of the main means of marking the frontiers of liberal government.

I begin with Shaftesbury in view of his location at the junction of two traditions. First, he was heir to the critiques of Hobbes advanced by the seventeenth-century natural rights theorist Richard Cumberland who disputed the ordering power of sovereignty by seeing civil society as being like a work of art. Regular and proportional in its functioning and constitution, but without having been legislated or produced by any political authority, the order of civil society testified, instead, to the 'invisible hand' of a divine providence. Second, as Josef Chytry (1989) has shown, Shaftesbury was also indebted to the Florentine tradition of civic humanism and thence to the neo-Roman conception of freedom. This was distinguished by the stress it placed on the responsibility of government to secure the conditions for freedom by providing political institutions in which citizens could take part in civic affairs free from relations of dependency or servitude that would topple them from virtue by making them susceptible to corruption or intimidation by others (Skinner 2002).

This Florentine legacy was reworked by Shaftesbury's concern, in the post-1688 'Whig settlement', to give a new definition and shape to the norms of gentlemanly culture constituted by codes of politeness. These codes, according to Lawrence Klein (1994), constituted an interactive conversational practice that served both to mark out distinctions within the body politic and to distribute authority in new ways in a society that was post-courtly, post-godly, pre-professional and pre-meritocratic. By constructing an intersubjective domain through the exchange of feelings and opinions via discourse on the arts and letters, the culture of politeness assumed differences of opinion while also operating as a mechanism for coordinating and reconciling those differences, allowing the self to be formed in dialogic relations with others. In its more demotic versions, this culture reflected the declining influence of the court and church and the rise of the new forms of urbanism associated with the development of the West End and of the spa cities with their network of institutions (coffee houses, clubs, gardens, promenades and theatres) that organised new forms of sociability between elements of the aristocracy, the landed gentry and mercantile and business interests. In such contexts the 'paradigm of politeness offered an alternative to the reliance on traditional authoritative institutions for ordering the discursive world, because it sought processes within the babble, diversity, and liberty of the new discursive world of the Town that would produce order and direction' (Klein 1994: 11–12). The model for Shaftesbury's conception of conversational practice was a more limited one related more closely to the position of the Whig gentry. Its ideal form, Klein argues, was more that of gentlemen conversing in a coach in a country house park. Here politeness had little to do with the multiple-stranded dialogism of town life but operated, instead, as a counter to the impolite and authoritarian practices of clerics and virtuosi.

However, it also operated as a counter to the Hobbesian account of sovereignty and its role in the production of social and political order. For Shaftesbury, polite discourse on the beautiful was to provide a basis for political authority that depended neither on divine right nor on Hobbesian might but which aimed, rather, for the governed to be 'all sharers (though at so far a distance from each other) in the government of themselves' (Shaftesbury, cited in Dowling

1996: 5). Polite discourse about questions of taste and judgement was to be translated into an inner mechanism of self-governance through what Poovey calls the surgical splitting of the self that Shaftesbury effected by translating the dialogical aspects of sociable conversation into a means through which the self conducts a dialogue with, and regulates, itself by bringing its many parts into harmony. Shaftesbury proposes, as a model for this splitting of the self, the dramatic soliloquy through which the character 'carries on the business of self-dissection' via an inner conversation with absent others, thereby becoming 'two distinct persons ... pupil and preceptor' (Shaftesbury 1999: 72) so that he is thereby able both to teach and to learn at the same time. The soliloquy thus constitutes a model of self-governance that is achieved through the internalisation of the mechanism of conversation with others.

This turned out, however, to be a mechanism through which individuals were to adjust themselves to an already ordered order. Although a strong opponent of the church, Shaftesbury believed in a divinely ordered cosmos governed by principles of mathematical proportion, which, however, the individual could come to apprehend directly and freely (rather than through the mediation of a priesthood) through the practice of judgement conceived as a disinterested contemplation of the manifestation of mathematical proportion in the orders of beauty. 'For Shaftesbury, then', as Poovey summarises the position, 'human subjectivity – the ground of the liberal governmentality – was formed in the image of mathematical order, and thus was naturally attracted to it. Society could be orderly ... because individuals ... wanted nothing more than to actualise the order they perceived in themselves' (Poovey 1998: 179).

There are a number of aspects to the role that disinterestedness plays in Shaftesbury's aesthetic. The one that Poovey refers to here concerns how Shaftesbury related the principles of disinterested observation, derived from the role that 'reliable witnesses' had played in establishing the truth claims of the empirical sciences (Shapin and Schaffer 1985), to aesthetic debate. A second aspect refers to his severance of aesthetic pleasure from the uses of possession or ownership (although, as Guyer [2005: 110–11] shows, not from all forms of utility) as a necessary condition for aesthetic judgement to be formed through conversational practice. The most significant aspect is, however, in common with the culture of civic humanism more generally (Barrell 1986, 1989 and 1992; Bohls 1993), his limitation of the capacity for disinterestedness to those (i) whose possession of landed property and ability to control their sexual desire freed them from dependence on others, (ii) who were entitled to bear arms, and (iii) whose liberal occupations gave them the time and intellectual capacity to have a care for the public weal free of self-interest. The liberal political community whose disinterested judgement, formed through its members' sociable and discursive relations to art, provided the basis for a new form of civic reason thus excluded 'those good rustics who have been bred remote from the formed societies of men or those plain artisans and people of lower rank who, living in cities and places of resort, have been necessitated however to follow mean employments and wanted the opportunity and the means to form themselves after the better models' (Shaftesbury 1999: 85).

Unsurprisingly this conception of disinterestedness was roundly criticised as both a creature and cover for the interests of the landed faction: Hogarth satirises it when he notes that 'it has ever been observ'd at all auctions of pictures, that the very worst painters sit as the most profound judges, and are trusted, I suppose, on account of their *disinterestedness*' (Hogarth 1997: 19). This and similar criticisms provide a template for what has been and remains, in the work of Bourdieu and the literature it has inspired, a criticism of the disinterestedness of the aesthetic as a mask for special interests, legitimating and perpetuating divisions of wealth and distinction. However, I want to take a different tack here by drawing on Jacques Rancière's approach to the aesthetic as a system for the distribution of the sensible, which operates less to legitimate economic and social inequalities than to produce and mark a political division: an unequal distribution of political and civic entitlements according to which only those whose position allows them to become judges of art are able to take part in government as both governors and governed (Rancière 2004 and 2004a). What matters from this perspective is not disinterestedness as a means of establishing a distance in economic and social space from those whose horizons are limited by dint of the constraints of material necessity but its role in producing a position in political space, which confers on those who can exercise command over and control of the self the capacity to direct the conduct of others.[4]

Shaftesbury's account of the space for dialogue with the self that he models on the soliloquy has been particularly important here. For it is a mechanism that has proved to be detachable from its particular anchorage in the landed gentry in Shaftesbury's work to form a part of more expansive conceptions of liberal government while simultaneously continuing to mark the frontiers – of class, gender and race – beyond which it could not be extended. As Vivienne Brown has noted, however, Shaftesbury frequently mixed his metaphors, often switching from the auditory associations of the soliloquy to describe how the self is split into two so that it might monitor and act on itself and to favour the hybrid auditory/spectatorial imagery of 'a vocal looking glass' to describe this mechanism (Brown 1994: 59; see also Brown 1997). There are also places where the soliloquy is described in exclusively spectatorial terms, as men become regular inspectors of themselves in acting as governors of themselves:

> And – what was of singular note in these magical glasses – it would happen that, by constant and long inspection, the parties accustomed to the practice, would acquire a peculiar speculative habit; so as virtually to carry about with them a sort of pocket-mirror, always ready, and in use.
>
> (Shaftesbury 1999: 87)

It is this mechanism, primarily in its spectatorial form, but still with traces of the auditory associations of the soliloquy, that Adam Smith develops in his account of 'the man within' in *The Theory of Moral Sentiments*. Working through norms of reciprocity – the mutual mirroring of the views of others – that Smith derives from Hume's account of natural sympathy, it is the mechanism through which men become governors of themselves:

> And, in the same manner, we either approve or disapprove of our own conduct, according as we feel that, when we place ourselves in the situation of another man, and view it, as it were, with his eyes and from his station, we either can or cannot enter into and sympathise with the sentiments and motives that influenced it. We can never survey our own sentiments and motives, we can never form any judgement concerning them; unless we remove ourselves, as it were, from our own natural situation, and endeavour to view them as at a certain distance from us....We endeavour to examine our own conduct as we imagine any other fair and impartial spectator would examine it. If, upon placing ourselves in his situation, we thoroughly enter into all the passions and motives which influenced it, we approve of it, by sympathy with the approbation of this supposed equitable judge. If otherwise, we enter into his disapprobation, and condemn it.
>
> (Smith 2002: 128–9)

The invocation of natural sympathy here transforms the specular mechanism that Smith had derived from Shaftesbury. In Shaftesbury's account, aesthetic judgement, while shaped by the subject's conversational interlocutors, is ultimately validated by the belief, as Caygill summarises it, (i) that there is a providence that disposes of all things into (ii) a beautiful order that (iii) gives to human beings the capacity to recognise and act in accordance with that order, which (iv) thus ensures a harmonious ordering of individual judgements and the general good without the need for state intervention (Caygill 1989: 46). By contrast, Vivienne Brown argues, in *The Theory of Moral Sentiment* 'truly moral outcomes are open; they are not rule-bound or obligatory but are the result of an open process of debate between the moral agent and the impartial spectator, in which the final outcome is neither predetermined nor legislated upon by a theological determinism' (Brown 1994: 64). Here, before the 'invisible hand' of *The Wealth of Nations*, the catoptric production of the unity of the manifold through a Hobbesian act of sovereignty is replaced by the emergence of order through reciprocal adjustments of conduct on the part of the members of civil society.[5] These are prompted by the 'sideways glances' that are produced by the system of mirrors, by means of which the members of civil society view their own conduct through the eyes of others.

However, as with Shaftesbury, this 'specular morality', as Poovey (1995) calls it, is reserved for those whose station in life equips them with the capacity to acquire it. Smith does not consider women as possible candidates for this role, and he discounts both savages and the mob, albeit on different grounds. The savage is driven by the harsh exigencies of necessity to exercise a Spartan discipline that affords neither the time nor the space to give or expect sympathy from those around him (Smith 2002: 239–43). His, then, is an iron-like self-command driven entirely by need. The mob, who, like Shaftesbury, Smith represents as still being dazzled by spectacle, are to be governed by leading them to obey sound moral codes routinely via the mechanism of habit rather than that of self-conscious assent acquired through the mechanisms of self-inspection. However, Smith is also clear that, even among the higher social classes, the capacity for an inner-directed morality is only available to 'a select, though, I am afraid but small party' whose capacity for 'studious and careful' observation allows them to discern that the path of the wise and virtuous is 'more exquisitely beautiful in its outline' and more worthy of emulation than is the 'gaudy and glittering' model of wealth and power that impresses itself 'upon the notice of every wandering eye' (Smith 2002: 73).

Autonomising the aesthetic, multiplying the uses of uselessness

Smith's appeal to aesthetic criteria as a means of sifting those who are most fully capable of exercising self-government, and thus the most qualified to be responsible for the government of others, echoes his earlier reference to the perfection embodied in works of art as an analogy for calibrating the degree to which individuals meet or fall short of the standards of self-command that good governance requires.

> It is in the same manner that we judge of the productions of all the arts which address themselves to the imagination. When a critic examines the work of any of the great masters in poetry or painting, he may sometimes examine it by an idea of perfection, in his own mind, which neither that nor any other human work will ever come up to; and as long as he compares it with this standard, he can see nothing in it but faults and imperfections. But when he comes to consider the rank which it ought to hold among other works of the same kind, he necessarily compares it with a very different standard, the common degree of excellence which is usually attained in this particular art; and when he judges of it by this new measure, it may often appear to deserve the highest applause, upon account of its approaching much nearer to perfection than the greater part of those works which can be brought into competition with it.
>
> (Smith 2002: 32)

The idea that is enunciated in the first of these standards of perfection – that is, the idea of a standard of perfection that is beyond the reach of any actual or potential work of art – might seem to anticipate Balzac's 1831 story *The Unknown Masterpiece* (Balzac 2001) in its expression of what Hans Belting calls the Romantic idea of 'art in the absolute – a state beyond the reach of every tangible art work' (Belting 2001: 11). Yet this is not so. Smith's formulation rather reflects the classical conception of the masterpiece in which, Belting argues, 'it was still possible to talk of art's perfection' (24). The 'principle of perfect art' that this involved – a principle that 'never materialised in a single work, but ... emerged from an imaginary synopsis of all existing historical schools and painters ... called for strict imitation, under the watchful eye of an all-powerful art criticism by official authorities' (22). By contrast, the regulative principle of the modern masterpiece – the idea of absolute art – is, as we saw in Chapter 3, one that renders the idea of any standard of perfection obsolete. It represents an impossible ideal, a necessarily unachievable horizon, which propelled the dynamism of an art system that had broken with the prescribed rules of the academies in favour of individualised forms of creativity whose accomplishments would always be incomplete in relation to the ideal of absolute art – art as its own end – that animated them. 'A *masterpiece* that was to be admired in a museum', Belting writes of this new conception of the masterpiece, 'contrary to the meaning of the term once used for a craftsman's presentation piece, now, paradoxically, reappeared as the free creation of a "genius"' (17).

However, I am less interested here in the institutional inscriptions of this form of canonicity than in the role played by Kant's conception of the relations between the ideas of beauty, genius and taste in auspicing a new kind of authority in place of the rule-bound expertise of the academies. Kant disposed of the latter by severing all questions concerning the judgement of taste from those concerning standards of perfection. Since 'a judgement of taste is an aesthetic judgement, i.e., a judgement that rests on subjective bases, and whose determining basis cannot be a concept and hence also cannot be the concept of a determinate purpose', it follows that 'in thinking of beauty, a formal subjective purposiveness, we are not at all thinking of a perfection in the object, an allegedly formal and yet also objective purposiveness' (Kant 1987: 74). His conception of genius as an exemplary form of originality, by contrast, serves as a provocation to, and model for, originality in others. 'Genius', he says, 'is a *talent* for producing something for which no determinate rule can be given, not a predisposition consisting of a skill for something that can be learned by following some rule or another; hence the foremost property of genius most be *originality*' (175). This provided the template for a new form of canonicity that Paul Guyer characterises as the 'locus of constant upheaval' (Guyer 2005: 251) rather than a stable canon identified in terms of definite norms of perfection. Kant again:

> Accordingly, the product of genius (as regards what is attributable to genius in it rather than to possible learning or academic instruction) is an example that is meant not to be imitated, but to be followed by another genius. (For in mere imitation the element of genius in the work – what constitutes its spirit – would be lost.) The other genius, who follows the example, is aroused by it to a feeling of his own originality, which allows him to exercise in art his freedom from the constraint of rules, and to do

so in such a way that art itself acquires a new rule by this, thus showing that the talent is exemplary.

(Kant 1987: 186–7)

If Kant's account of genius is central to the distinction he proposes between the mechanical and fine arts, that distinction is in turn crucial to the role that his account of the aesthetic plays in energising the internal economy of the subject through the imaginative free play of the relations between the faculties that it engenders. He accordingly develops his account of genius as much with regard to its consequences for the subject – for the consumer of fine art – as for the production of a new from of canonicity. In contrast to his earlier assessments of the potentially invigorating effects of the forms of self-reflection that might be induced by a poem, a novel or a comedy,[6] Kant condemns the merely agreeable effects of the mechanical arts for their failure to unsettle the subject in ways that would promote critical forms of self-reflection. The originality of genius, however, generates a corresponding degree of indeterminacy – and so scope for freedom – in the reception of the work of fine art. Such a work, Guyer argues, must be the result of an intentional activity to be recognised and classified as art by the audience. However, the artist's understanding of his own work cannot dictate the audience's response without eliminating the free play that is the hallmark of the aesthetic. The aesthetic conception of the work of art thus rules out the possibility of 'complete uniformity in the synchronic or diachronic reception of a truly successful work of art' (Guyer 2005: 255).

This does not mean, though, that Kant abandons the art gallery visitor entirely to his or her own devices. To the contrary, while wishing to rescue the free play of the aesthetic from the subsumptive rationality of Wolff's philosopher king, the very purpose of Kant's critique of judgement was to harness the free self-activity of the subject to the process of culture understood as a process of human self-making guided and directed by reason. Freedom, Kant writes, chastising the pleasure of the sublime when its celebration of the liberality of the imagination remains at the level of mere play, is only properly understood as 'a law-governed *task*' (Kant 1987: 128). Judgement is not a matter of bringing the subject's taste into compliance with an externally imposed standard of taste. Rather, as a practice of freedom, it is a capacity that is developed through use. It is equally the exercise of this capacity within civil society that produces the standard with which individual judgements will be progressively aligned. This standard, the *sensus communis*, represents the collective reason of mankind that is produced via the historical anamnesis of tradition. As such, it is enacted through culture – through, that is, productive activity and communication with others – as the process through which the finality of nature and that of humanity are brought into accord with one another in a just proportionality that is the outcome of their interactions.

If this construal of the relations between judgement and culture removed the production and reception of art from the superintendence of guilds and academies with their fixed rules and definite standards of perfection, their place was taken by an authority of a new kind, the philosopher-aesthetician who apportions to different activities their roles in respect to this process. With unusual bluntness, Kant subordinates the untrammelled originality of genius to the superintendence of the power of judgement so far as its role in the progress of culture is concerned:

> Taste, like the power of judgement in general, consists in the disciplining (or training) of genius. It severely clips its wings, and makes it civilised, or polished; but at the same time it gives it guidance as to how far and over what it may spread while still remaining purposive. It introduces clarity and order into a wealth of thought, and hence makes the ideas durable, fit for approval that is both lasting and universal, and [hence] fit for being followed by others and fit for an ever advancing culture.
>
> (188)

The relations that civic humanism posited between the freedom of the subject and the uselessness of the aesthetic are reworked by Kant to produce a new space for the aesthetic in which the exercise of judgement was brought under a new kind of authority, which, rather than prescribing a rule for judgement, sought to guide it so as to secure the ends of culture. This conception of the autonomy of the aesthetic produced a new historical–discursive space in which the 'uses of uselessness' multiplied as the newly constituted 'authorities of culture' sought to guide and direct the practice of freedom by harnessing judgements of taste to a variety of social purposes as necessary way-stages for the realisation of culture's goals.

Jacques Rancière's elaboration of the aesthetic regime of art inaugurated by Kant's aesthetic makes this relationship between the autonomy of the aesthetic regime of art and the multiplication of the spheres of art's practical application unusually clear. 'The aesthetic regime of the arts', he writes, '… frees [art] from any specific rule, from any hierarchy of the arts, subject matter, and genres. … It simultaneously establishes the autonomy of art and the identity of its forms with the forms that life uses to shape itself' (Rancière 2004a: 23). By freeing art from any specific rule, such as that of particular artistic hierarchies or genres, the aesthetic regime of art effects a redistribution of the sensible to the extent that it promises both a new life for art and a new life for individuals and the community. If, Rancière argues, this secures the autonomy of art from its earlier conscription in service of social and occupational hierarchies, it also grounds that autonomy 'to the extent that it connects it to the hope of "changing life"' (Rancière 2002: 134). Rather than interpreting autonomy and heteronomy as opposing principles defined in a relationship of simple antagonism to one another, Rancière argues that the aesthetic regime of art generates a series of different "emplotments" of the relations between autonomy and heteronomy. Flaubert's *l'art pour l'art*, the educative mission of museums and libraries, early industrial workers' search

for freedom through literature, Adorno's negative aesthetics and the com-
mercialisation of art as a means of bridging the division between the art of
the beautiful and the art of living: these are all different 'emplotments' of the
possible courses of action that are generated by the 'and' that derives from
'the same knot binding together autonomy and heteronomy' (134). They all
work within the space of separate-but-connected that this knot effects, but
they work within the tensions these coordinates establish to different purpose
and effect.

Far from standing opposed to instrumentalisation, this conception of art's
autonomy requires it. A "muddying" of the relations between art and life is
precisely what the aesthetic regime of art makes possible in its concern to
connect art to the task of changing life. The forms that this muddying has
taken have varied. In Britain, rather than generating a tradition of formalised
aesthetic theories, as it did in Germany, Kant's influence was mediated via
Coleridge and the Romantics and thence, via its connections to the question
of character that preoccupied liberal political thought, from John Stuart Mill
through to Matthew Arnold, to debates about the role of education in shaping
self-governing forms of individuality (Roberts 2004).

The influence of civic humanist conceptions, particularly those derived
from the work of Joshua Reynolds, remained influential in shaping the ethos
and practice of public art galleries through to the mid-nineteenth century
(Barlow and Trodd 2000). Post-Romantic conceptions of the aesthetic similarly informed a range of interventions into the uses of art, literature and
music that were concerned, precisely, with 'making culture' so as to 'organise
freedom' and thereby 'change society' by enfolding new constituencies within
the dynamics of self-reform that was the Kantian bequest.[7] Drawing on
George Gissing's contention that 'nothing in this world is more useful than
the beautiful' (cited in Maltz 2006: 7–8), Diana Maltz has traced the impact
of this bequest on a range of nineteenth-century social projects. However, she
focuses particularly on those late-nineteenth-century forms of 'missionary
aestheticism' in which the radical forms of autonomy ascribed to art by the
British Aesthetic Movement auspiced the activities of whole armies of
voluntary and philanthropic workers, for whom art provided the means for a
range of interventions into the conditions of working-class existence, in their
endeavours to bring art to bear on the task of changing life.

We can hear the echoes of these reformatory endeavours well into the
twentieth century. We can hear them, for example, in Jonathan Rose's account
of the role that a literary or art education had for generations of working-class
students who expressed their appreciation of its benefits precisely in terms of
the dual relations of disinterestedness and freedom, and their influence on
both self and social reform. Take, for example, the following comment from
the 1936 *Learn and Live* survey of Workers Education Association students:

The giving of prominence to things of the mind and spirit and the
encouraging of an attitude of mind which places man first and his

economic function second; freedom from commercialism; disinterestedness. All of which, I believe, go to stimulate the student to social service.

(Rose 2001: 284)

And we can hear them in John Carey's suggestion that reading literature might serve as an effective antidote to binge drinking (Carey 2005). In his polemic *What Good are the Arts?* Carey discusses a report in *The Times* in which two 15-year-old schoolgirls construe their binge drinking as an attempt to escape the boredom of life in a small Gloucestershire village. Connecting this to the decline of the public library, Carey prescribes reading good literature as a mind-enhancing and life-changing escape from boredom in contrast to the merely temporary escape offered by drugs, drink and antidepressants. As such, he presents exposure to good literature as an alternative to the attempt to tackle binge drinking by punitive or corrective measures. In doing so, he proposes a contemporary version of nineteenth-century dreams of luring the working man away from the pub and into the library or art gallery as a means of combating drunkenness and all the social ills (wife beating, promiscuity and unchecked population growth) that accompanied it (Bennett 2000).

These might seem to be two quite different 'takes' on the relations between freedom and culture that Kant proposes. In the first, connected to the self-emancipation of the working class, culture, as the escape from necessity, represents the free pursuit of knowledge as an end in itself while, in the second, culture is interpreted as a mechanism of government that works through the means of self-regulation it makes possible. But these two conceptions of the relations between culture and freedom are closely linked. The history of the relations between adult education in Britain and the development of literary education (Lee 2003) show how closely the former was modelled on the latter as a space of free expression in which, via their responses to the literary text, students were encouraged to express their selves in specially controlled settings (the tutorial). This made their thoughts and feelings open to correction and revision in the light of the guidance offered by an experienced reader (the tutor) and their fellow students. 'Its essential characteristic', Albert Mansbridge said of the Workers Education Association Tutorial Class, 'is freedom. Each student is a teacher, each teacher is a student' (cited in Rose 2002: 276). It was, indeed, precisely this extension of the use of literature from its previous association with elite forms of self-cultivation to the population more generally via the public schooling system and adult education that placed the literary class at the heart of a new machinery of self-government (Hunter 1988). Although I doubt that it was what he had in mind, the pliability of the aesthetic in these regards is aptly summarized by Georges Batailles' remark that it is only when the aesthetic work has been separated from the history of its particular uses, and so no longer serves any specific purpose, that it acquires the paradoxical value of 'absolute usefulness' (Batailles, cited in Hollier and Ollman 1991: 13).

7 Guided freedom

Aesthetics, tutelage and the interpretation of art

In one of his many discussions of the concept, Theodor Adorno defined aesthetics as a practice of interpretation and commentary that aims to produce a critical and self-reflexive form of individuality that 'stands free of any guardian' (Adorno 1963: 281).

It is a formulation that aptly captures how the understanding of aesthetics as a practice of freedom has been developed in the slipstream of Kant's account of the beautiful as 'what is presented without concepts as the object of a *universal liking*' (Kant 1987: 53). Freed from the constraint of any determinate concept, aesthetic judgement was thereby also putatively freed from the sway of moral or political authorities as the interpreters of such concepts. By thus disentangling aesthetic from conceptual judgement, Kant produced the aesthetic as a zone of activity – a practice of the self – that could be conducted independently of any tutelage to such external authorities, a space of freedom. Foucault addressed these aspects of Kant's legacy – discussed in more detail in the previous chapter – on a number of occasions, notably in his commentary on Kant's 1784 'What is Enlightenment?' (Foucault 2007a) and in his essay on the aesthetics of existence (Foucault 1989). He returned to the same concerns in 1982–3 in his course of lectures on *The Government of Self and Others* where, confessing that Kant's essay on Enlightenment had become 'something of a blazon, a fetish for me' (Foucault 2010: 7), he offers a probing discussion of its evasions and contradictions. These centre on Kant's conception of Enlightenment as '*man's way out from his self-incurred tutelage*' to acquire the capacity 'to make use [of his understanding; MF] without direction from another' (Kant, cited in Foucault 2010: 26).

There are two aspects of Foucault's discussion I want to highlight here. The first concerns Kant's negative assessment of men's capacity to escape from their tutelage through their own action and volition, and, equally, the inability of exemplary individuals to lead or guide others to the autonomy that they had attained without, thereby, exercising an authority over them that would, in the very process of 'freeing' them, bind them into a new form of tutelage. The second concerns the conclusion Kant draws from this: that Enlightenment consists in a redistribution of the relations between the government of self and others affected by a new arrangement of the relations between private and public reasoning. This depended, Foucault notes, on 'an ingenious little trick' through which Kant reverses the normal meanings of these words. Private reasoning, as Foucault summarises it, relates to 'our public activity as functionaries, when we are components of a society or government whose principles and objectives are those of the collective good' (35). In such cases we are 'he [Kant] says, just "parts of a machine" ... placed in a given spot, [with] a precise role to play, with other parts of the machine having to play different roles' (35). Public reasoning, by contrast, relates to 'the use we make of our understanding and our faculties inasmuch as we place ourselves in a universal element in which we can figure as a universal subject' (36). We constitute ourselves as such subjects, Foucault continues, 'when as rational beings we address all other rational beings' (36). Kant's argument, though, is not for the complete displacement of private reasoning, or of the forms of tutelage it involves, by public reasoning, but a proper apportionment of the relations between the two, such that the latter – in which there is no relationship of tutelage to any authority – orders the spheres in which the former is exercised.

However, as Foucault notes, Kant's essay offers no account of the process through which this condition is to be arrived at. This is rather the accomplishment of his later (1790) *Critique of Judgement.* By freeing the judgement of beauty from its subordination to concepts, this opened up a space – individually within the architecture of the person, and collectively within the historical architecture of humanity – in which judgement is freely exercised in a manner that paves the way for the production of a space of universality in which the principles of pubic reasoning might hold sway. The aesthetic thus operates as a technology of the self of a particular kind. Its definition as a form of sensory pleasure that cannot be brought under any concept, Thomas Osborne (1998) argues, opens up the sphere of art – for artists, critics and audiences – as one characterised by a struggle for autonomy from the imposition of any particular code of conduct derived from moral, political or scientific authorities. The work of art, freed from the restrictions of canonical authorities, becomes the zone of a potentially limitless but, at the same time, unachievable freedom. Together with the struggles of artists who seek constantly to renovate this capacity of art in order to reach beyond its historically determinate forms and so open up new avenues of expression into a limitless prospect of free inventiveness, the work of art serves as the template for an endless refashioning of the self that is not brought under the sway of any particular moral code.

Such aesthetic conceptions of art have, as we have seen, been deployed in varied ways in mediating the relations between art and its publics. Osborne's formulations underline their role producing a historically specific form of engagement with art that there are good reasons for valuing. But only, I want to argue, on certain conditions and within certain limits. These have ultimately to do with the unsustainable contention that the aesthetic does indeed, as Adorno contended, produce a relationship in which the self 'stands free of any guardian'. This fails to recognise the respects in which the space of the aesthetic itself produces distinctive forms of tutelage that induct individuals into certain practices of 'guided freedom' that are subject to the direction of a distinctive kind of authority. It is also an unusually combative form of authority, pitching itself against all other forms of authority, and in particular against those associated with the empirical disciplines and their use in calculating the civic effects of different kinds of art-public encounters. This is the authority that the philosopher claims in setting himself up as an 'authority of freedom' whose role is to superintend the capacity of freedom that arises from the subject's encounter with the aesthetic. This is, though, a form of authority that occludes its own role, and that of the apparatuses in which it is enacted, in organising the coordinates of the aesthetic encounter.

I pursue these concerns via a critical assessment of Jacques Rancière's account of the relationships between the aesthetic regime of art, freedom and politics. There is, as I have indicated, much to value in Rancière's approach to the plural social inscriptions of artistic practices within the aesthetic regime of art. There are, though, good reasons for considerable caution regarding the position that he himself takes up within this regime in his advocacy of that particular social inscription of the aesthetic that he characterises as meta-politics. Consideration of these will throw useful light on the general position of contemporary critical aesthetics. This is so for three reasons. First, Rancière's conception of metapolitics constitutes the most comprehensive and influential attempt to suture back into place the severance of the social and artistic critiques that fuelled the protests of 1968 (Boltanski and Chiapello 2007). Second, Rancière's conception of the relationship between the aesthetic regime of art and the project of metapolitics provides an unusually clear insight into the operations of the Kantian armature that underpins assessments of the aesthetic as a space for the exercise of a putatively universal judgement over and against civic forms of reasoning. Third, Rancière is unusual in the degree to which his advocacy of the aesthetic is urged by combating the authority of other forms of expertise. This is particularly true of his criticisms of the empirical disciplines, especially sociology, and the place these occupy in his account of the relations between police, as a practice of social ordering, and his conception of politics as an aesthetically motivated practice of freedom. I therefore take this as my starting point.

Art and the division of occupations

Let us go back to Kant for a moment. By arguing that the aesthetic, as a distinctive mode of perception, could not be brought under any concept, Kant made it immune to any explanation that might be derived from any science. While this was initially the basis of his critique of Alexander Baumgarten's ambition to establish a science of the aesthetic, it has subsequently provided the ground for the rebuttal of attempts to account for the aesthetic in sociological terms. As that which pleases without a concept, the aesthetic cannot be reformulated in any terms except its own without distorting its most essential and distinguishing properties. The very attempt to develop a sociology of art, it is argued, subjects the work of art to a form of expertise that traduces the properties of its object. As Adorno states the position:

> Once art has been identified as a social fact, there is a growing sense of superiority on the part of sociologists who wish to do nothing more than to control art. Hewing to an ideal of positivistic value-neutrality and objectivity, they flatter themselves that their knowledge is superior to what they discredit as a collection of subjective standpoints in art and aesthetics. Efforts like these must be combated. They quietly seek to enforce the primacy of the administered world over art, whereas art wishes to be left alone and acts up against total socialisation.
>
> (Adorno 1984: 355)

This is the tack Rancière takes in his various disputations with sociology, principally as represented by Pierre Bourdieu. Rancière sets out his stall clearly enough in *The Philosopher and His Poor* (1983), where his characterisation of Bourdieu as 'the sociologist king' places him in the same position as both Baumgarten and Christian Wolff's philosopher-king: that is, as someone who seeks to lord it over the aesthetic by bringing it under a system of concepts. Rancière thus questions the very nature of Bourdieu's enterprise in *Distinction* (1984). By asking questions exploring people's liking for and knowledge of different kinds of music, Bourdieu's social surveys aim, Rancière says, to 'judge musical tastes *without having anyone hear music*' (Rancière 2004: 187). By transforming a test of musical taste into one of knowledge – a move incompatible with Kant's definition of beauty as that which pleases without a concept – 'the sociologist has solved the problem without even tackling it' (187). The artwork as such disappears from the analysis, dispersed across the different fields and the struggles comprising them that are the social scientist's toolbox of concepts. And the artwork, as the locus for a practice of freedom, is transformed into a prop of class domination as the relations between freedom and necessity are stretched across a set of polarised class tastes. Professionals and managers, in demonstrating their disinterested appreciation of beauty as 'that which pleases without a concept', transform art – and the art gallery – into a means of performing and symbolising their social distance from the working classes whose tastes remain mired in the lack of freedom constituted by their choice of the necessary.

I do not wish to make light of these criticisms. Indeed, as I have argued elsewhere (Bennett 2007c), they probe significant limitations of the methodological protocols that guide Bourdieu's analysis in *Distinction*. However, it is less clear that Rancière achieves his main purpose: namely, to establish that what he calls the aesthetic regime of art, introduced by Kant's aesthetic, disconnected art from the hierarchical division of occupations that had characterised the earlier forms of its social existence so as to effect a redistribution of the sensible that suspends such social conditioning of aesthetic tastes and competences. This is difficult to square with the wealth of statistical material that bears on these matters. Let me make the point via a visual illustration of the findings of a recent survey conducted, in the tradition of *Distinction*, of the social distribution of cultural practices in Britain (Figure 7.1).[1]

This 'map' offers a visual plotting of the relations between cultural practices. Each symbol represents the statistical mean point for the members of the survey who either do or do not like or take part in the activity indicated. The circles refer to cultural participation – to things that people do or do not do. A zero indicates nil or very low levels of participation; the numbers 1 and 2 indicate occasional and high levels of participation respectively. The

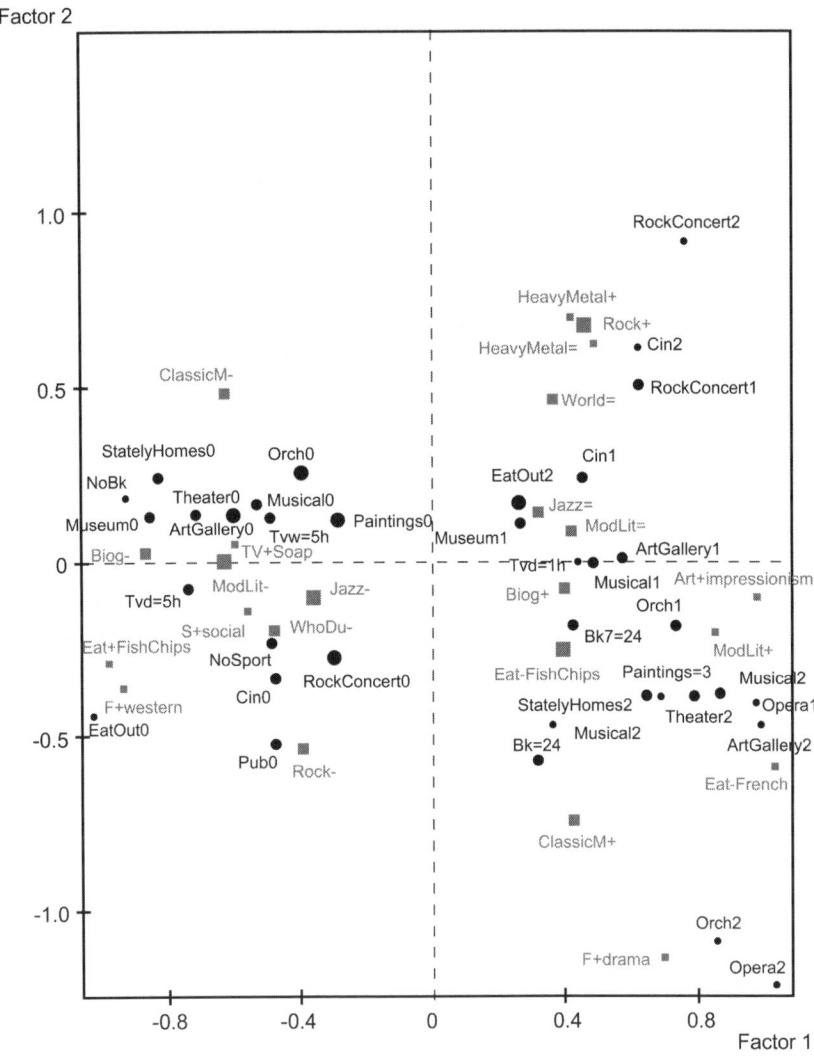

Figure 7.1 The space of British lifestyles, 2003–2004: primary axis of differentiation.
Source: *Culture, Class, Distinction*, Bennett *et al.*, Routledge, 2009

squares refer to tastes: a plus sign indicates liking, a minus sign indicates dislikes and an equals sign indicates neutrality. The size of the symbol indicates the number of people engaged in, or not engaged in, liking, or disliking, the activity concerned. The key point, finally, concerns the degree of proximity or distance between the practices that are represented in these ways. The greater the degree of proximity, the greater the degree of overlap between the members of the sample engaged in or liking the activities concerned. Where the distance is greater, the likelihood of such overlap is correspondingly lower.

As can be seen, high levels of participation in, and a positive liking for, activities and genres conventionally associated with the aesthetic cluster to the right-hand side of this space: art gallery visitation, going to the theatre and opera, going to orchestral concerts, liking classical music, liking impressionism, owning a number of paintings and liking modern literature. Conversely, the left-hand side of the space is defined by zero or very low levels of involvement in, or liking for, these activities. The social distribution of these tastes and forms of participation across classes stands out sharply when we map the occupational class positions of the members of this sample into this space (Figure 7.2). Participation and preferences are lowest among all sections of the working classes, particularly among routine and semi-routine workers, and highest among the middle classes, particularly professionals, and employers and managers in large organisations.

Of course, this map only gives us half the picture. It tells us solely about the distribution of those tastes and practices that are statistically most distinct

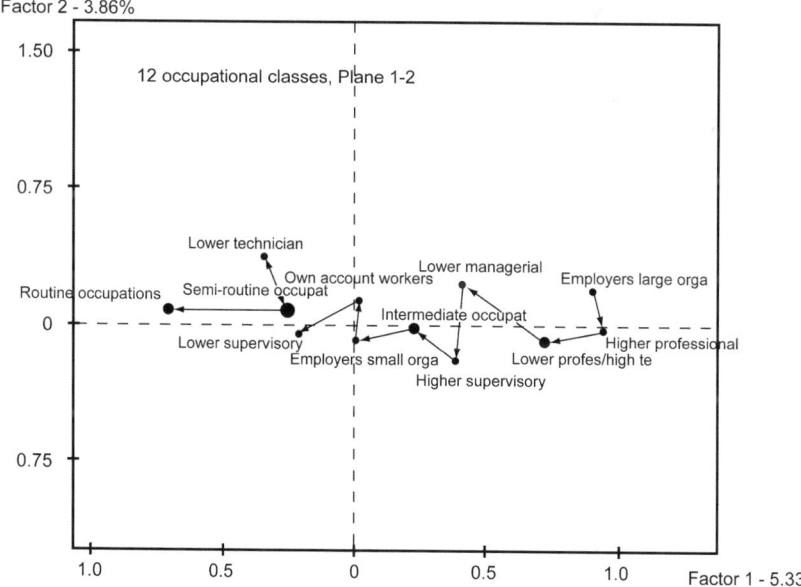

Figure 7.2 The distribution of class positions across the space of lifestyles.
Source: *Culture, Class, Distinction*, Bennett *et al.*, Routledge, 2009

from each other. There is thus a visually 'suppressed middle' comprising those areas of cultural practice – many of them connected with film, television, music and sport but also visual art (landscape paintings, for example) – in which the tastes and preferences of different classes intermingle. While these were included in the survey, they do not show up in a procedure designed to detect statistically significant differences in tastes rather than commonalities. Nor does this evidence in and of itself dispose of the objections Rancière registers in relation to sociological surveys of this kind. It does not gainsay his contention that an aesthetic sensibility slips out from the hold of conventional hierarchies of the arts to constitute an aspect of the affective and sensory investments in more popular art forms.[2] To the contrary, a closer look at particular aspects of this study shows that this is, indeed, the case: such investments are likely to be more intense in relation to rock music, for example, than they are in relation to classical music, especially among younger managers and professionals.[3]

While important, such qualifications do not detract from the general conclusion that the relationships of different sections of the population to artistic practices and institutions continue to be marked by their relations to the divisions between occupations. Indeed, in this study, artistic practices are more sharply divided by class than by any other social variable. Rancière's objection to Bourdieu, however, is less an empirical one than one voiced in the name of the emancipatory possibilities of Kant's aesthetics. Kant, Rancière says, refuses 'the absolutisation of the gap between working-class "nature" and the "culture" of the elite' that he sees in Bourdieu's account of the relations between the bourgeois principles of 'pure taste' and the working-class choice of the necessary. He seeks instead 'the anticipation of the perceptible equality to come, of the *humanity* that will be the joint surpassing of the culture of the dominant and the culture of Rousseauist nature' (Rancière 2004: 198). His contention, then, is that the aesthetic is a social force that might lead to the overcoming of the divisions between occupations.

There is, in truth, not as much clear water between Rancière and Bourdieu on this matter as Rancière would like to imagine. For Bourdieu's account of the aesthetic is also subtended by a similar post-Kantian historical narrative, but one that is given a 'sociological twist'. It is, as I argue in greater detail in Chapter 9, a narrative that works in terms of the relations between fields, the position of the collective intellectual within these, and the relations between such intellectuals, the state and the education system, through which class differences in relation to the aesthetic are eventually to be eliminated. The question that is of more immediate relevance, however, concerns the kind of authority Rancière deploys against that of the sociologist and the empirical disciplines more generally. Hayden White has described Rancière's style of reasoning as 'more aphoristic, even oracular, than demonstrative or argumentative', noting that this 'makes it virtually impossible to submit what he asserts about anything whatsoever to any test of falsifiability on the basis of evidence' (White 1992: xi, xviii). I shall, in what follows, build on this insight by arguing that the place accorded the aesthetic in Rancière's work constitutes a species of 'guided freedom' in which freedom is brought under the direction of a form of prophecy that derives its authority from the secularised form of Christian eschatology that is the Kantian legacy.

Secular oracles

Michel Foucault's remarks concerning Kant's 'ingenious little trick' in reversing the ordinary understanding of the meanings of 'public' and 'private' suggest a helpful route into these questions. For the manner in which Rancière inserts aesthetics into the relations between politics and police depends on his own equally ingenious trick of defining these against the grain of their received usage.[4] He reinterprets the conventional view of politics as 'the set of procedures whereby the aggregation and consent of collectivities is achieved, the organisation of powers, the distribution of places and roles, and the systems for legitimising this distribution' as, instead, the defining attributes of police (Rancière 1999: 28). Rancière borrows the latter concept from Foucault (1991), and follows him in interpreting police as a set of general processes of social ordering as distinct from the 'petty police' (Rancière 1999: 28) or police force. However, his usage departs from Foucault's in two key respects. First, it loses the historical specificity that characterises Foucault's account of police as a historically intermediate form of *raison d'État* that paves the way for the (relative) transition from sovereign to governmental power. Police, for Rancière, describes the ordering of the proportions between the rights, entitlements and the distribution of rewards that accrue to different sections of the population in the Greek polis just as much as in contemporary France. Second, however, Rancière narrows the definition of the term in limiting the exercise of police to a particular set of ordering functions. Police distribute bodies across social and political space by instituting a particular ordering of the field of the perceptible. Rancière defines this field as

> an order of bodies that defines the allocation of ways of doing, ways of being, and ways of saying, and sees that those bodies are assigned by name to a particular place and task; it is an order of the visible and sayable that sees that a particular activity is visible and another is not, that this speech is understood as discourse and another as noise.

(29)

Police, crucially, distributes rights to speak by ordering distinctions within the division of occupations between those who, in the terms of Aristotle's *Politics*, have the capacity of voice – a capacity of expressing pain or pleasure that men share with animals – but not that of speech: that is, the ability to offer an account, and to be taken into account, in discourse concerning the just and the unjust.

Politics, by contrast, is whatever disrupts the orderings of police by asserting a right to speech that undermines the distinction between voice, or noise,
and speech, and does so in the name of the equality of speaking beings:

> What is usually lumped together under the name of political history or
> political science in fact stems more often than not from other mechanisms
> concerned with holding on to the exercise of majesty, the curacy of divinity,
> the command of armies, and the management of interests. Politics only
> occurs when these mechanisms are stopped in their tracks by the effect of
> a presupposition that is totally foreign to them yet without which none
> of them could ultimately function: the presupposition of the equality
> of anyone and everyone, or the paradoxical effectiveness of the sheer
> contingency of any order.
>
> (17)

Politics, on this definition, does not play a part in the distributional struggles
about the allocation of rights and rewards across the division of humanity
into different occupations. It intrudes into such orderings of the social, governed
by principles of mathematical proportion, the assertion of the equality of
speaking beings. It is an intervention produced by those who, previously of no
account, lacking any political or civic status, assert their right to be seen and
to be heard, and thus to be taken into account. They do so, however, not as
one part among others in the distributional stakes of 'equity politics', but
as an enunciation of the community – a community no longer rent by the
divisions between occupations – that is yet to come but which, simply by
articulating the demand for it, becomes virtual, a component in the makeup
of the present, in the Deleuzian sense. Politics does not occur when 'social
groups have entered into battle over their divergent interests' but when particular social forces (the Athenian demos, the proletariat), hitherto denied a
voice, assert a demand that dissolves the proportional logic of those forms of
social ordering governed by the principles of police in 'the sheer name of
equality between anyone and everyone by means of which classes disconnect
and politics occurs' (18). Politics is, in short, metapolitics, a set of discursive
interventions into and with the principles of police that, while eschewing the
position of a pure outside to power or one of absolute negation,[5] nonetheless
indicts existing forms of political struggle over distributional issues in the
name of a community to come that will displace such concerns.[6]

The extended use of these two terms – police and politics – is central to
Rancière's method that, in spite of his denial of anything quite so systematic
(Rancière 2009a), has a definite logic, one that depends on his ability to
attribute equivalent effects to the speech acts of historically distinctive actors
in radically different historical circumstances. He does, however, accord these
two terms a more historically specific usage in his account of the relationships
between, on the one hand, modern forms of police and the empirical disciplines,
particularly sociology, and, on the other, the aesthetic regime of art and the

form of politics it generates. With regard to the former, Rancière objects to Bourdieu's work as typifying the structural logic exhibited by sociology more generally in its concern to establish systematic correlations between particular social positions on the one hand and particular aesthetic and epistemological dispositions on the other.[7] That Bourdieu, or the sociologist more generally, might argue for a reformist programme that reorganises the relations between social positions and dispositions by redistributing cultural capital remains, for Rancière, a position marked by the logic of police. It is a practice conducted 'from above' by the sociologist in collaboration (however critically) with the state, and on the basis of a form of expertise that lays claim to a knowledge of the connections between social positions and dispositions that eludes the occupants of those positions.

Rancière's conception of the aesthetic cuts into these relations between sociology and police in a distinctive way. In registering his distance from Walter Benjamin's thesis of the fascist 'aestheticisation of politics', Rancière argues that aesthetics has played a longer-term and more foundational role in providing the distinctive basis and rationale for politics in its modern form:

> The modern emergence of aesthetics as an autonomous discourse determining an autonomous division of the perceptible is the emergence of an evaluation of the perceptible that is distinct from any judgement about the use to which it is put; and which accordingly defines a world of virtual community – of community demanded – superimposed on the world of commands and lots that gives everything a use.
>
> (Rancière 1999: 57)

Rancière's understanding of the aesthetic here is not as a theory of the specificity of art or of the beautiful, or of sensibility, but rather as 'an historically determined concept which designates a specific regime of visibility and intelligibility of art, which is inscribed in a reconfiguration of the categories of sensible experience and its interpretation' (Rancière 2006: 1). The disconnection of art from concepts and from use, from knowledge and from desire, effected by Kant, cuts into the hierarchical orderings of genres, and of their uses and publics, associated with earlier regimes of art, to (potentially) float art free from its earlier inscriptions within the order of occupations. Art thereby becomes a disordering force with the potential to uncouple such inscriptions:

> A well-ordered society would like the bodies which compose it to have the perceptions, sensations and thoughts which correspond to them. Now this correspondence is perpetually disturbed. There are words and discourses which freely circulate, without master, and which divert bodies from their destinations, engaging them in movements in the neighbourhood of certain words: people, liberty, equality, etc. There are spectacles which disassociate the gaze from the hand and transform the worker into an aesthete.
>
> (9)

The reference to spectacles here alludes to a passage from Kant's *Critique of Judgement* on which Rancière places considerable weight. It is a passage in which Kant explores what is called for in order for a palace to be considered beautiful rather than, as the role it had played in Adam Smith's *Theory of Moral Sentiments* (Smith 2002: 61–74), an object of stupefaction for the mob – a thing, Kant says, 'merely to be gaped at' (Kant 1987: 45) – or, as in Kant's summary of Rousseau's estimation, an occasion for rebuking 'the vanity of the great who spend the people's sweat on such superfluous things' (46). Kant's answer – or the aspect of his answer that Rancière notes – is that a palace can only be judged beautiful provided that the subject adopts a stance of disinterestedness in relation to it, focusing on its form rather than its uses.[8] This serves as the basis for a lengthy refutation of Bourdieu, which relies on Rancière's interpretation of the testimony of a woodworker who, in 1848, in a moment of leisurely reflection, recorded in his journal a description of his relationship to the scene of his work, which disconnects that scene from its functioning as a site of his own exploitation, or as a testimony to the wealth and power of its owner, to transform it into an object of disinterested plea-sure. In thus effecting 'a disjunction between an *occupation* and the *aptitudes* which correspond to it' (Rancière 2006: 5) the woodworker – or so Rancière would have us believe – brings the house of sociology tumbling down. On the site thus vacated Rancière proceeds to construct his own edifice by interpreting the woodworker's pleasure as a doubling of his identity. While still a worker, Rancière's diarist also assumes the identity of a proletarian understood not as a sociological category but as a subject who escapes the limitations of an assigned class position, located outside the political community, to intervene in the affairs of that community in the name of a community that is yet to come.

A woodworker who tells us a tale that suggests the Kantian aesthetic slips out of the class limits to which the sociologist would confine it, and who then serves as an oracle for what is to come: the procedure is one that runs throughout Rancière's work. The figures he invokes are always artisanal – woodworkers, cobblers, tailors. They are always invoked to demonstrate a misalignment between aesthetic aptitude and social position; that misalignment is always attributed to the worker's use of his leisure time in ways that run counter to the effects of his class positioning in the relations of production;[9] and it is always interpreted as beckoning toward a redistribution of the sensible that will create a political community beyond the schismatic effects of the division of labour.

It is pointless to object that such singular exceptions do not constitute an adequate refutation of the probabilistic logic on which sociological accounts of the distribution of aptitudes rest; and fruitless to observe that 1848, or 1832,[10] do not tell us much about where and how such exceptions occur in the twenty-first century, or what significance we might attach to them when the historical links that existed between Kantianism and early labour political organisations are considerably more attenuated.[11] Nor is much purpose served by pointing to the atypicality of artisanal and craft-based forms of employment, then and now, compared to the conditions of factory work, call centres and the global export of sweatshop labour. If lodging such objections would be pointless this is because they are all arguments that would require Rancière to submit his work to the authority of the empirical disciplines – to history[12] and sociology – and thus to the procedures of police. This would run counter to his purpose of elaborating a conception of politics as a practice of freedom founded on the equality of speaking subjects and modelled on the aesthetic disposition. And pointless, too, because his vignettes of shoemakers, carpenters and schoolmasters[13] serve as the occasions for a practice of oracular divination that, by interpreting these figures as anticipations of a community that is to come, brings the practice of freedom under the guidance of another kind of authority: that of a secularised Christianity.

To suggest that Rancière invokes a form of authority might seem quixotic given his association with the *révoltes logiques* collective (Ross 1991), and his commitment to the introduction of disagreement into the exercise of all forms of social ordering. The mantle of the aesthetic that he lays a claim to, however, is one that generates a form of authority that depends precisely on a gesture of disavowal as the precondition for its exercise.[14] It is a form of authority produced by the displaced form of Christian metaphysics that Kant's aesthetic organises. I draw here on the work of Ian Hunter who argues that, by loading the concept of 'humanity' with all of the attributes previously attributed to god, and projecting its completion as the outcome of a secular historical process, Kant secured a continuing role for a secularised and rationalised form of Christian metaphysics (Hunter 2001). The production of a difference between humanity as an ideal yet to come and actual human beings provided a new set of discursive co-ordinates within which a new regimen of the self might be formed and distributed under the tutelage of the Kantian philosopher-aesthetician. As heir to the Pietistic tradition and its substitution of new forms of self-rule in lieu of subordination to priestly systems of morality,[15] Kant constructed a new form of intellectual and cultural authority on the part of those able to close the gap between the two humanities by a disinterested purification of their human-sensible nature that detached it from any specific ends. The aesthetic, within this schema, constituted a practice that, although not guided by concepts, was conducted under the tutelage of the persona – the philosopher aesthetician, the genius, the work of art itself – exemplifying the ideal forms of comportment it aspired to.

This space – historically sculpted out of, but remaining attached to, a particular religious metaphysic – is the space that Rancière's project of metapolitics occupies. This proposes a framing of the reception of art in which a still operative Christian metaphysic provides the coordinates for grooming a corps of paradoxical freedom fighters who are subject to a kind of authority that operates above and beyond the empirical realm of mundane politics, the state and civic reasoning in the name of a humanity yet to come. The exercise of this authority depends on particular techniques of self-grooming that place the aesthetic consumption of art in a different register: the cultivation, for example, of what Rancière describes as the attributes of 'non-possession and passivity', or of a demeanour that seeks to overcome the 'dissensual intervention of political subjects' by looking beyond 'the appearances of democracy and of the forms of the State to the infra-scene of underground movements and the concrete energies that comprise them' (Rancière 2009: 33, 35). It is in the name of this authority that Rancière takes issue with any and all forms of theory and interpretation that might concern themselves with the subject's relations to art from the point of view of any social purpose it might serve, of how the civic benefits of particular kinds of art practices might be calculated, or of how relations to art institutions might be equalised.

Producing and distributing freedom

Rancière is not alone in advancing arguments of this kind. Peter Osborne (2006) and Simon During (2010) have argued for similar positions, and in both cases for reasons that absolve art from the messy business of policy, administration, bureaucracy and civics. It is, however, Rancière who offers the most elaborate version of this argument, and whose work constitutes a rallying point for a project of politics, derived from the aesthetic, as a practice of the equality of speaking subjects who, to recall Adorno's phrase, 'stand free of any guardian'. My purpose, in this and the preceding chapter, has been to sketch a genealogy for this position by tracing some key moments in the development of the discourse of the aesthetic that confer on its spokesmen a particular kind of authority. In developing his account of the relations between Kant's conception of humanity and Christian metaphysics, Hunter stresses the role this played in Kant's struggle to undermine the contending tradition of civil philosophy. By placing philosophy in the service of empirical forms of reasoning that would temper the clash between different religions by aspiring to be indifferent between them, the latter tradition constituted a 'rival Enlightenment'. In lieu of this, Kant produced a space in which a particular cadre of intellectuals, occupying a distinctive position within the German university system, pro-duced a historically new form of authority based on a capacity to decipher the essential truths of human existence via transcendental forms of reasoning. In doing so, he also produced a new space within the self as the inner stage on which this authority could be exercised as a form of tutelage that effaced itself in the process of orchestrating a freedom that seemed free of any guide.

Viewed in this light, the battle lines Rancière draws between aesthetics and sociology are a historical re-run of those that Kant drew, through a combative intellectual practice that was far from disinterested, between his own project of transcendental critique and empirically inclined forms of civic reasoning. I do not, though, want to validate my criticisms of Rancière by appealing to the forms of sociological authority he disputes. My purpose is rather to locate his concerns within the problem space of liberal government as a set of prac-tices, developing in tandem with the principles of security, which regulate conduct by producing, organising and distributing freedoms of various kinds.

As we saw in Chapter 2, freedom, in this account, is the outcome of a new governmental rationality that, by placing limits on the exercise of state power, produces the conditions for new zones of freedom to emerge. Freedom is not a given but something that has to be produced and organised:

> The formula of liberalism is not 'be free'. Liberalism formulates simply the following: I am going to produce what you need to be free. I am going to see to it that you are free to be free.
>
> (Foucault 2008: 63)

This production of the conditions in which (some) individuals are free to be free is the work of intellectual and cultural authorities of various kinds. In undertaking this work, such authorities also distribute freedom. It is within the space of this new governmental rationality that modern aesthetics develop as a new technology for governing through freedom; that is, for regulating conduct through the new kinds of freedom it makes up via its discourse on art, freedoms that it apportions differentially to different parts of the social body. Bourdieu and Rancière are both equally, albeit different, products of this technology. They both mobilise the forms of authority and techniques of veridification associated with the space of the aesthetic with a view to effecting a redistribution of the freedom that is its product. In Bourdieu's case, this is a redistribution of freedom along class lines that is to be effected by the equalisation of access to the aesthetic disposition. In Rancière's, it is a matter of the redistribution of the sensible effected by the aesthetic regime of art as a resource for the project of a metapolitics pitted against the principles and procedures of police. These positions by no means exhaust the possibilities that are generated by aesthetics understood as a liberal technology for the production and distribution of freedom. To the contrary, this technology has been deployed across a range of pedagogical, economic and political practices. Indeed, as I argued in Chapter 6, this is aptly captured by Rancière's account of the varied 'emplotments' of the relations between art and life that have been generated by the aesthetic regime of art depending on whether these relations have been fashioned by Marxist, Romantic, Hegelian or Schillerian aesthetics, and on the properties of the different apparatuses or *dispositifs* through which such emplotments have been enacted (Rancière 2009: 33–4).

These are, however, matters that Rancière is able to put to one side by virtue of the further 'ingenious little trick' that validates the emancipatory potential he attributes to the aesthetic. For his interpretation of the aesthetic as founding a politics that disturbs the relations between voice and speech, between mere animal noise and a stake in the community, depends solely on his assessment of the benign potential that can be deciphered in the embrace of an aesthetic disposition by specific fractions of the skilled sections of the nineteenth-century working classes. Yet this can withstand the burden Rancière places on it only if it is assumed that the redistribution of the sensible across the divisions between occupations that he attributes to the aesthetic has a paradigmatic status for its functioning across all social divisions. This is to discount the respects in which the aesthetic regime of art has also been a party to a parallel series of 'redistributions of the insensible' through which varied populations have been civically and politically incapacitated, placed beyond the limits of freedom by being reduced, precisely, to mere animal noise. If we look beyond the Western aesthetic regime of art to consider its colonial encounters with other art systems, we find that it has operated precisely as a system for turning those who were previously of some account into people of no account, of people who had a part into people with no part. The Kantian conception of the aesthetic has served as a means of undermining non-Christian forms of religious authority by promoting a disinterested interest in form that would detach the passions and the senses from the grip of idolatry whose power, in denying the independence of judgement required by the tenets of classical liberalism, barred the colonised from any claim to a stake in the political community (Roy 2006). It also informed a long history of anthropological tests designed to assess the distribution of the sensible across racial divisions that, by denying the 'primitive' an adequately developed capacity for sensory discrimination, justified extreme forms of racial oppression.[16]

It is histories of these kinds that are now most at stake in the relations between 'interpretation, theory and the encounter',[17] both in the art gallery and in relation to the more general production and circulation of art in our contemporary 'globalised' world.[18] As such, the issues they raise are ones that Rancière is unable to come to terms with, partly because he has rendered the discursive space from which he speaks free of the contradictions that have characterised it historically, and partly because he has no means of engaging with the relations between the aesthetic, art practices and social divisions except for those defined in terms of occupations. Where can he stand in relation to Aboriginal artists who, in wishing to limit the force exerted on their practice by the aesthetic regime of art, have strenuously resisted interpreting their art as an expression of free and innovative creativity?[19] And how, given his remarks on the 'ethical regime of images' (Rancière 2009: 28), in which works of art are perceived and judged in accordance with their relations to the norms and standards of divinity, can he address the complex issues that are now posed by the place of art practices within multicultural polities where, like it or not, the art gallery's publics are now religiously diverse in ways that go beyond sectional divisions within Christianity? Since, as we have seen, Rancière lives in a metaphysical glasshouse that is of a distinctively Christian construction, he is hardly in a position to throw stones at those for whom art stands as the representative of divinities other than his own.

However, it is not my purpose to suggest that the forms of 'guided freedom' that have been produced in association with the aesthetic regime of art – understood, as Rancière proposes, as a regime of art practices whose reception is mediated by post-Kantian philosophical aesthetics – are to be discounted as ideological delusions. Thomas Osborne (2008), in a later work than the one I referred to at the start of this chapter, offers a more qualified assessment of the legacy of post-Kantian aesthetics for the role it has played in cultivating a particular kind of ethics, one that, rather than seeking to instil any particular set of moral norms, promotes an open-ended questioning of all such norms. This constitutes what Osborne calls an ethos of 'educationality' – a stirring up of things, prompting the reader into self-reliant judgement – as distinct from 'pedagogy', understood as teaching a particular moral system. The model of 'educationality' is not that of an exemplar in pursuit of followers; but of an exemplar for subjects of judgement who will follow their own path. Yet, while valuing this legacy, Osborne also wants to place limits on the reasons for, and respects in which, it should be valued. For the capacity it promotes, he says, is one that, since it allows us to work on ourselves to ward off the force of particular moral systems that inhibit and limit creative thought, is also unable to provide any positive guidance as to who we are or should strive to be, or what to do. Its role, he argues, has thus to be understood as ancillary in relation to other kinds of undertakings: moral, political, scientific, civic or, indeed, utopian.

But this, of course, is to bring the aesthetic under a concept and thus to deprive it of the autonomy on which claims to an essential alignment between aesthetics, freedom and critique have depended. However, since, as we have seen, this autonomy is illusory – the effect of a very particular kind of freedom produced by the interested organisation of an intersection between Christian metaphysics, aesthetic discourse and art practices – this is nothing to get upset about. In 2008 the McMaster report recommended that the focus of UK government arts funding should shift away from 'social good' objectives (reaching minority ethnic groups, for example, or the 'socially excluded') – practices that, in Ranciere's terms, subordinate the aesthetic to police – to focus instead on artistic excellence without having to translate this into definable social or political benefits (McMaster 2008). Rónán McDonald, endorsing this proposal, argued that 'it is precisely by being "useless" that art can be most useful to society', and urged the need for the public sphere 'to rediscover the language [of uselessness] to engage with this paradox' (McDonald 2008: 39). My purpose has been to engage with this challenge by disclosing some of the uses to which aesthetic discourses of uselessness have been put. I have done so, however, precisely with a view to undercutting the ground that informs the opposition between the aesthetic, on the one hand, and that of the procedures of bureaucracy or calculations of social or civic utility on the other.

I have thus suggested that, when viewed as a liberal technology for governing via a particular form of 'guided freedom', aesthetics might be best understood as a historically and culturally distinctive form of 'process ethics' that is more concerned to induct individuals into particular ways of shaping their conduct via particular procedures of self-inspection than it is to prescribe particular moral codes. When considered in this light, however, it is best interpreted not as a singular exception to other ethical practices but as one among other 'process ethics' that have developed over the same period and that can equally claim a provenance in the complex and divided history of the Enlightenment. Bureaucracy, rather than being construed as art's other – as part of a police/ politics polarity – thus emerges from the pen of Max Weber as precisely a parallel form of 'process ethics' embodying a commitment to disinterested forms of impersonality that detach the duty of office from any commitment to a particular set of moral or political ends.[20] The history of the civic mediation of art practices via the use of empirical instruments to assess how such practices might form a part of programmes aimed at the amelioration of conflict in multicultural policies can equally claim an inheritance in the 'process ethics' of the tradition of civic philosophy that Kant opposed.[21]

Let us, then, thank Rancière for his provocations. But let us, at the same time, recognise the historical freight that is sedimented in the form of authority he mobilises, and the limitations – ethical, political and analytical – that this brings in its tow.

Part 4

Habit and the architecture of the person

Inter-text 3

The aesthetic, I have argued, constitutes a distinctive kind of authority that, like any other, works through the specific kinds of tutelage it organises. I have sought both to provide a rough genealogy for, and to illustrate, the qualities of the guided forms of freedom this gives rise to by probing the historical and discursive underpinnings of Jacques Rancière's work. Jane Bennett alludes to another aspect of these in recording that, when asked whether an animal, plant or drug might disturb the orders of police, 'Rancière said no: he did not want to extend the concept of the political that far; nonhumans do not qualify as participants in a demos; the disruption effect must be accompanied by the desire to engage in reasoned discourse' (Bennett 2010: 106). His position in this respect is classically Kantian in its conception of culture as a unique realm of free human agency. This is not, though, as I noted in Chapter 1, solely a dividing line between the human and the nonhuman; it is also a dividing line between different humanities, a dividing line that is typically drawn around the place that habit is accorded within the makeups of different forms of personhood.

It is to these questions that I now turn in Part 4 where I examine the roles played by both anthropology and aesthetics in these regards. I look first, in Chapter 8, at the concept of habit associated with the late-nineteenth and early-twentieth century anthropological doctrine of survivals. I do so with a view to identifying the light this can throw on how and why Australian Aborigines came to be thought of as unimprovable and therefore no longer susceptible to the civilising effects that it had earlier been thought would result from their exposure to culture. My concerns here will serve as an adjunct to my discussion in Chapter 5. It adds to these a concern with the ways in which different 'architectures of the person' inform the governmental deployment of cultural knowledges. I then, in Chapter 9, look at Bourdieu's understanding of the relations between habit and habitus. I trace, first, the legacy of the anthropological doctrine of survivals in Bourdieu's discussion of the organisation of what he calls 'archaic habitus'. I then discuss the essentially Kantian historical schema underlying his account of the role that collective intellectuals are to play in distilling the potential for freedom and distributing this capacity more broadly to encompass those whose habitus are shaped more by the routine repetitions of habit. My purpose in doing so is to extend the disciplinary reach of my discussion by showing how Bourdieu's sociology can be approached as a liberal discipline that operates through the differential distribution of the capacities for freedom that it produces.

8 Habit, instinct, survivals
Repetition, history, biopower

In 1844, Lord Stanley, Secretary of State of the Colonial Office, wrote to Sir George Gipps, the colonial governor of New South Wales, regarding a report that Gipps had forwarded to him from Captain G. Gray. Drawing on his experience as the commander of an expedition into the interior of Australia, Gray had dwelt on the lacklustre results of all the attempts that had so far been made to civilise the Aborigines. Stanley acknowledged that it seemed 'impossible any longer to deny' that such attempts 'have been unavailing; that no real progress has yet been effected, and that there is not reasonable ground to expect from them greater success in the future' (cited Anderson 2007: 120). Yet he was reluctant to accept the conclusion that followed from this. Noting that he could not admit 'that with respect to them alone the doctrines of Christianity must be inoperative, and the advantages of civilisation incommunicable', he declined to believe that Aborigines '[are] incapable of improvement, and that their extinction before the advance of the white settler is a necessity which it is impossible to control' (120–21).

Kay Anderson argues that Stanley's equivocations over this matter were symptomatic of a moment when Australian colonial discourses were poised between two options. On the one hand, both Christian salvationist discourses and the secular progressivism of Enlightenment stadial theory allowed – indeed, urged – that the Australian Aborigine might be improved. Set against these views, increasingly influential somatic conceptions of race rooted racial divisions ineradicably in the body and, thereby, removed Aborigines from both the Christian time of salvation and the progressive time of civilisation, placing them instead in the dead-end time of extinction as the inevitable losers in the struggle for existence with a superior race. This somatisation of race initially took the form of polygenetic accounts of racial divisions that called into question both Christian and Enlightenment accounts of human unity. While Darwin's account of evolution opened up a space in which Aborigines might be enfolded within civilising programmes by denying that racial differences were innate or constituted unbridgeable gaps, Anderson suggests that subsequent developments in anthropology placed Aborigines beyond the reach of such programmes by consigning them to the newly historicised twilight zone between nature and culture represented by the category of prehistory. As survivals of the past in the present, Aborigines presented the difficulty not of being innately different but of being too far away in time. Still on the cusp of the journey from nature into culture, they had simply too far to travel across the eons of evolutionary time separating them from the properly historical time of their colonisers before the imperatives of racial competition resulted in their elimination.

I have no quarrels with this account that, indeed, resonates with many aspects of my argument in Chapter 4. My purpose here, however, is to argue that the distinctive dynamics that connected a belief in the 'unimprovability' of Aborigines and the doctrine of survivals in the late-nineteenth and early-twentieth centuries depended on the ways in which the relationships between habit and instinct were reconfigured in the context of post-Darwinian social, political and anthropological thought. For this pluralised and historicised the concept of innateness in ways that reordered its relations to race. This argument also serves as a vehicle for a broader purpose: to shade and qualify the role that has been attributed to habit within liberal forms of government in the post-Foucauldian literature on governmentality (see, for example, Mehta 1997; Valverde 1996; White 2005). This has largely been concerned with habit as a mechanism distinguishing where the assumption that individuals are to be governed through their capacities for freedom should apply and where, instead, more coercive forms of rule should be brought into play. Where behaviour has become so habituated through frequent repetition that it trespasses on the capacity for the will, guided by reflexive judgement, to be freely exercised, the shutters have been drawn on liberal strategies of rule in favour of reinforcing the mechanisms of habit as an automated form of self-rule.

This argument has proved of considerable value in highlighting the wide range of exclusions – of race, class, age and gender – through which liberal government has been constituted. Its chief limitation is that it fails to take account of the different places that habit has occupied within the architectures of the person associated with different discourses and strategies for organising 'the conduct of conduct'. By 'architectures of the person' I have in mind what Nikolas Rose (1996a) characterises as a historically mutable set of 'spaces, cavities, relations, divisions' that are produced by the infolding of diverse ways of partitioning the self and working on its varied parts that are proposed by different authorities. I develop this argument by examining how post-Darwinian social, political and anthropological thought shifted the place that habit occupied within the architecture of the person that had been proposed by classical liberalism by refashioning its relationship to custom on the one hand and instinct on the other. The consequence of this, I argue, was a distinctive habit–instinct nexus that inscribed the governance of 'unimprovable' Aborigines in a specific form of biopower. I shall, in this connection, revisit the work of Baldwin Spencer to examine the respects in which his racialisation of Aboriginality was informed by the distinctive configuration of the relations between habit and instinct that emerged from these post-Darwinian trajectories of classical liberalism. First, though, to provide a contrapuntal historical backdrop to these concerns, I look at the role played by the concept of habit in earlier moments in the development of English liberal political thought.

Habit, custom and the will: a virtuous cycle

Patrick Joyce's account of the anxieties that clustered around the role of habit in mid-nineteenth-century British conceptions of liberal government provides a good point of departure for these concerns. Joyce attributes these anxieties to the position that habit occupied within an architecture of the person wherein, by mediating the relations between desire and compulsion, it problematised the subject as the locus of both stasis and change: 'habits are ingrained in nature, but can none the less be broken by the power of the will' (Joyce 2003: 118). If both personal and social development required that the force of habit be broken, this could only be with a view to installing another set of habits in its place. 'Habit', as Joyce puts it, 'must counter habit' (120). The exercise of the will must both pit itself against habit and instil a new set of routines through which conduct is regulated if the ideal of a constantly self-renovating personhood that is capable of both transforming and stabilising itself is to be realised. This, in turn, is necessary in order that society might continue to progress through the free activity of its subjects. The case Joyce has most in mind is that of John Stuart Mill's account of the logic of the moral sciences – first published in 1843 – that reconciles freedom and necessity by attributing to the will the capacity to remake habits and, by thus reshaping the self and asserting mastery over habitual forms of conduct, to exercise a capacity for moral freedom (Mill 1967). In Mill's account, as Melanie White glosses it, 'the presence of the "will" and an established foundation of good habits generate the dispositions necessary for the responsible exercise of freedom', while it is the role of character to judge which habits further and which impede its own moral dynamic and thence to begin 'a slow process of developing counter-habits and routines in order to reinforce the will, and also to reflect changes in the moral expression of one's character' (White 2005: 482).

Yet it is notable that at no point does Mill's discussion of habit go beyond the dialectic of will and habit to open up questions concerning the relations between habit and instinct, understood as either a natural foundation for habitual dispositions or as a hereditary mechanism for their transmission across generations. In this respect, his account echoes Locke's assessment that education, fashion and custom prevail over innate dispositions in accounting for habitual regularities of conduct. Mill's view of the part played by reflexive judgement in reviewing the hold of custom similarly echoes Locke's account of the role played by moments of 'uneasiness' in opening up to inspection the customs that, through repetition, have come to be installed as the habits that constitute a particular 'relish of the mind'.[1] Here, then, in Locke's account, habit operates as a vital mechanism in a virtuous cycle through which conduct is endlessly shaped, and reshaped.

This stands in contrast to Kant for whom habit, understood as instinct, stood in a vertiginous opposition to the will that, in accordance with the Kantian project of purifying subjectivity to free it from all material contingencies, had little habitual about it. In *Anthropology from a Pragmatic Point of View*, Kant distinguishes physical anthropology's concern with 'what nature makes of the human being' from a pragmatic orientation toward anthropology concerned with 'the investigation of what he as a free acting being makes of himself, or can and should make of himself' (Kant 2006: xiii). As such, he attributes habit wholly to natural or quasi-natural forms of conduct that, since they are driven by necessity, are devoid of moral significance. Since it 'is a physical inner necessitation to proceed in the same manner that one has proceeded until now', Kant argues, habit 'deprives even good actions of their moral worth because it impairs the freedom of the mind' (40). For Kant this association of habit with instinct aligns it with nature. Habit, to recall the passage I quoted in Chapter 1, arouses disgust because 'here one is led *instinctively* by the rule of habituation, exactly like another (non-human) nature, and so runs the risk of falling into one and the same class with the beast' (40). The only exception that is admitted to the rule that 'all habits are reprehensible' (40) is where they testify to the power of intentionality versus nature, as in the adoption of routine culinary habits to offset the effects of old age. For the rest, far from being inscribed as a mechanism in a virtuous circle of conduct formation, habit is clasped together with instinct as a couplet through which the power of nature as necessity works and to which the power of culture – understood as the capacity for free self-shaping – stands opposed. 'The animal creature he sets up as a foil to the human being', Sankar Muthu argues of Kant, 'is instinctively driven. The movement from animality to humanity is one toward freedom and culture' (Muthu 2003: 128).

It is this aspect of Kant's work that informs Mill's later essay 'On Liberty' (1859). Unlike his earlier discussion that focused on the relations between habit and the will, Mill's discussion here is organised in terms of the contrast between custom and character:

> A person whose desires and impulses are his own – are the expression of his own nature, as it has been developed and modified by his own culture – is said to have a character. One whose desires and impulses are not his own, has no character, no more than a steam-engine has a character.
>
> (Mill 1969: 74–5)

Where character is an inoperative force, however, this is because conduct is subject to the despotism of custom rather than the grip of habit. Although there are often areas of overlap between them, the two concepts are not identical. Custom, as Colin Campbell (1996) notes, may, as in the case of *sutteeism*, consist of singular rather than frequently repeated acts, just as the reasons for taking part in such acts may rest on conscious volition rather than – as the stock definition of habit – mechanical, unthinking repetition. In Mill's case, the despotism of custom is sometimes attributed to such a mechanism. He attributes conformity to custom among some peoples to their lack of any faculty except that of 'the ape-like one of imitation' (Mill 1969: 73). This is not, however, the organising core of Mill's account. If the spirits of liberty, progress and improvement are the attributes of character that stand opposed to custom, these can only flourish where they are supported by the political conditions of democracy that – through the mechanism of discussion – allows individual variation to become an active force in social life. 'I have said', Mill writes, 'that it is important to give the freest scope possible to uncustomary things, in order that it may in time appear which of these are fit to be converted into customs' (83). Conversely, the despotism of custom prevails wherever the mechanisms of discussion are underdeveloped or held in check. It is this aspect of the character/custom opposition that serves as the basis for Mill's account of the distinction between societies with and without history in the sense that Koselleck (2002) gives to this term: that is, the expectation that the future will be different from both the present and the past as a result of the changes initiated by self-conscious subjects acting within developmental time. Here again, then, where character serves as a principle that can call custom to account, habit, custom and the will interact as parts of a virtuous cycle of character formation, but one that develops along a progressive historical trajectory.

However, if history is impossible where these conditions do not apply, this is because, for different reasons in different historical circumstances, character and custom have locked in on one another in vicious, self-enclosing cycles of immobility. Mill thus interprets Asiatic societies as societies that, while once historically dynamic, have since exited from history through the enforcement of custom associated with 'Oriental despotism'. By contrast, he construes primitive societies as ones that have never entered history either because they are societies in which 'the race itself may be considered as in its nonage' (Mill 1969: 15), or because they are 'anterior to the time when mankind have become capable of being improved by free and equal discussion' (16).

These, then, are some of the ways in which the concept of habit informed the development of early modern liberal political thought. Its role in this regard, however, varied depending on the place it occupied in relation to adjacent concepts within different architectures of the person that laid out different relations of internal action of the self on self (will/habit; culture/instinct; character/custom) as one of the mechanisms through which liberal government operates. It is against this background that I look next at how the place of character within the architecture of the person mutated in relation to what Stefan Collini (1979 and 1971; see also Hawkins, 1997) calls the 'historicisation of character'. This played a key role, in the last quarter of the nineteenth century, in the transition from the earlier laissez-faire orientation of classical liberalism to the formulations of the new liberalism that envisaged a more interventionist role for the state, in particular, in aiding the development of character. This was, however, no longer a character system organised in terms of either an opposition or a virtuous cycle between will and habit, culture and instinct, or character and custom. Rather it laid out the person as a series of historicised, developmental gradations between custom, habit and instinct – that is, more in the form of a slope than an opposition – and interpreted instinct not as a pure nature opposed to culture but as an accumulating stock of conscious actions passed on into the automated forms of instinct via the mediatory roles of habit and inheritance.

The main intellectual development prompting this revision of the earlier character system of classical liberalism was Darwin's *Origin*. Published in the same year as Mill's *On Liberty*, this prompted a succession of attempts – on the part of Walter Bagehot (1873), Henry Maudsley (1902) and Lloyd Morgan (1896), for example – to account for how the forms of conduct acquired by habit in one generation could be passed on to the next as a set of inherited instincts by being deposited in the nervous system or some equivalent quasi-physical mechanism. This was, of course, very much a case of 'creative treason' that, as Laura Otis notes, owed less to Darwin, who by and large resisted the view that characteristics acquired by one generation could be inherited by the next, than it did to Baptiste Lamarck's account of the inheritance of acquired characteristics and, later, to Ernst Haeckel's biogenetic 'law' that ontogeny recapitulates phylogeny (Otis 1994: 3–10). More fundamentally, perhaps, the view that living beings are shaped by their interactions with their environment depends on a Lamarckian conception of the relations between the organism and its milieu in contrast to the emphasis Darwin placed on the struggle between different forms of life as the chief mechanism of variation (see Canguilhem 2008: 103–5). Nonetheless, the result was a decisive refashioning of the architecture of the person that, as Otis summarises it, introduced a new element into this architecture – that of 'organic memory', which 'placed the past *in* the individual, *in* the body, *in* the nervous system' (Otis 1994: 3) – while also laying out the person as a part of developmental sequences in which conscious and unconscious processes, the social environment and nature, interacted in new ways. It was an architecture within which 'memory and heredity, habit and instinct' operated 'as points on a continuum' leading to a 'steady accumulation' of competencies across generations, and which meant that the body could be read as 'a record, a palimpsest, perhaps, of its interaction with its environment, in its own lifetime, in its grandparents' lifetimes, and in the lifetimes of its distant ancestors' (6).

It is the place accorded habit within such historicist revisions of character by the late-nineteenth-century generation of social, sociobiological and anthropological theorists that especially concerns me here. My interest centres on the role they played in fashioning one of the more peculiar forms of liberal modernity associated with imperial Britain in the new terms of intelligibility they proposed for the 'unimprovability' of Aborigines and their consequences for the development of new forms of biopolitical administration.

From habit to instinct: somatic accumulation, evolution, history

In her preface to the English translation of Félix Ravaisson's *Of Habit*, Catherine Malabou locates its concerns at the junction of two philosophical traditions. The first, following a line from Aristotle through Hegel to Bergson – and, indeed Max Weber too – treats habit as a constitutive aspect of human existence: that is, as a permanent disposition and a virtue in stabilising conduct. The second, operating in terms of the mind–body dualisms that run from Descartes to Kant, interprets habit as pure negativity: 'the disease of repetition that threatens the freshness of thought and stifles the voice … of the categorical imperative' (Malabou 2008: vii). Malabou argues that Raivaisson's text mediates the relations between these two traditions by interpreting the stabilities produced by the repetitive mechanisms of habit as the precondition for the acquisition of an aptitude for change through which living beings are able to take part in the production of an open-ended future.

A key aspect of Ravaisson's argument here concerns his account of the relations between habit and instinct. In contrast to Kant, who places both of these on the side of nature in opposition to culture and the will, Ravaisson places habit between the will and instinct, interpreting it as the mechanism that translates actions initiated by the will into a 'second order' set of instincts through which 'primitive instinct' is transformed into an accumulating set of competencies. 'Habit', as Ravaisson puts it, 'transforms voluntary movements into instinctive movements' (Ravaisson 2008: 59). This posits an architecture of the person in which 'habit is the dividing line, or the middle term, between will and nature; but it is a moving middle term, a dividing line that is always moving, and which advances by an imperceptible progress from one extremity to the other' (59). It is not, however, only the dispositions of the individual that are affected in this way. It is through the linking mechanism of habit that nature itself is gradually transformed. 'In descending gradually from the clearest regions of consciousness', Ravaisson argues, 'habit carries with it light from those regions into the depths and dark night of nature' (59). The result is an ascending slope, without any abrupt transitions or dualistic oppositions of a Kantian type, through which all forms of life – from the will or motive activity to the simplest forms of life – are connected via the mechanism of habit.

Ravaisson's work drew on contemporary developments in physiology, albeit taking issue with those accounts that construed habit as purely an effect of motor mechanisms, contending that these failed to take account of the tendency given to habit by the lively force arising from the adaptive relations between the sense organs and the environment.[2] Similar concerns informed the post-Darwinian debates in the life sciences that provided a materially grounded alternative to the antinomies of Kantian philosophy while, at the same time, providing a space in which the emergence of will could be accounted for in terms of the relations between the organism and the environment. This was true of the development of the relations between biological and social evolutionism in Britain. In *The Principles of Psychology*, for example, Herbert Spencer accounts for the differentiation of reason and the will from the instincts not as a set of constitutively different faculties but as the outcomes of evolutionary processes of differentiation in which habit and – as an addition to Ravaisson's formulations – memory mediate the relations between reason, will and instinct. The individual organism, Spencer argues, responds to changes in its environment that it experiences as external shocks; frequent recurrence of the same shocks produce corresponding changes in the internal structure and dispositions of the organism; such repeated changes in dispositional behaviour lead to progressively more complex divisions in the organisation of the nervous system. If Spencer attributes the development of reflexes and instincts, and the higher faculties of will and reason, to this same general process, his subscription to a Lamarckian conception of the relations between milieu and organism allows for the transgenerational accumulation of competencies as a set of hereditable instincts.

The relations between 'conscious memory' and 'organic memory' (Otis borrows the term from Spencer) are central to this process. Conscious memory comes into play when the connections between a particular set of psychic states induced by changes in the milieu are no longer coordinated through the automatic mechanism of habit; and it passes away when such coordination once again becomes automatic by being passed on as a part of an accumulated instinctual inheritance that is transmitted to the next generation via organic memory. Here, Spencer remains faithful to the assumptions of Locke's empirical psychology while simultaneously recasting them. True, there are no innate faculties or ideas prior to experience, but this does not rule out the possibility of there being historical forms of innateness that are the somatic accumulation of the successive experiences of past generations that have come to be coded into the body as a set of compound instincts. And it is only this accumulating legacy of past experience that opens up the space and the time within which the higher faculties of reason and the will might emerge and be exercised. There is no break here between habit and the will; just a seamless transition: 'And this, the cessation of automatic action and the dawn of volition, are one and the same thing' (Spencer 1996: 614).

Henry Maudsley, whose writing on habit and the will played a significant role in late-nineteenth- and early-twentieth-century liberal thought (Valverde 1998), similarly stresses that there is 'no break or pause in the ascent from monad to man' (Maudsley 1902: 37). Simple reflex actions depend on 'a nervous machinery formed and fitted through remote ages now to act automatically' (37), whereas acquired reflex actions are subject to gradual formation through repeated practice, and eventually becoming automatic. The exercise of the latter is an art the individual learns for himself whereas the performance of the former is 'a function which has been learnt for him in a dateless past and he now inherits ready-made' (38). The will is merely 'the present culmination of organic evolution' (47) and, as such, it is the result of 'the same process at work now by virtue of which in the remote past the habits of prehistoric ancestors have become the instinctive and reflex faculties of today' (226–7).

But this is true only for some races as Maudsley goes on to differentiate races in terms of the depths of their inheritance of the somatic accumulation of the experience of earlier generations: the deeper the inheritance, the further the race has progressed. This leads him to suggest that habit might serve as a mechanism that will eventually 'perfect a rational and moral nature of the human species' by bringing the habits of less-developed races under the influence of more civilised ones. However, he immediately closes the door on this prospect, protesting 'how puerile and pernicious a practice it is to attempt to force the habits of one level of civilisation on people who are on a lower level, especially on those who are on a level of barbarism' (231). The problem here, given an architecture of the person laid out as a set of dispositions linearly connected to one another along an evolutionary trajectory, is one of sequence. How can the habits of those with fully developed somatic inheritances be grafted onto those for whom inheritance remains at a prehistoric level? It was in response to this problem – a problem produced by the doctrine of survivals developed during the interval that separates Maudsley's and Spencer's texts – that new, biopolitical terms of reference were brought to bear on the question of the Aborigine's capacity for improvement.

Exiting history: somatic and cultural 'flat-lining' and the logic of biopower

Let me recap. My purpose so far has been to consider the different roles that have been accorded habit depending how it has been distinguished from or aligned with other aspects of conduct in the architectures of personhood associated with different tendencies in British liberal social and political thought. However, these are not always so clearly distinguishable in practice. To the contrary, elements of different traditions were quite frequently in play in debates concerning the relations between habit and the regulation of conduct. Their implications for the 'unimprovability' of Aborigines were consequently framed in different ways at different points in time, even by the same person. This was true of Baldwin Spencer who initially construed the conservatism he attributed to the Arunta[3] of Central Australia in the terms proposed by Mill's account of the opposition between the despotism of custom and the democratic principle of discussion as the chief mechanism through which variation is introduced into a polity:

> As among all savage tribes the Australian native is bound hand and foot by custom. What his fathers did before him he must do. If during the performance of a ceremony his ancestors painted a white line across the forehead, that line he must paint. Any infringement of custom, within certain limitations, is visited with sure and often severe punishment. At the same time, rigidly conservative as the native is, it is yet possible for change to be introduced.
>
> (Spencer and Gillen 1899: 25)

To account for how such limited kinds of change might come about, Spencer invokes the principle of discussion. However, he does so in a way that explains how change can occur (it is prompted by the discussions that take place when different local groups meet) but at the same time be constrained within definite limits (these discussions are not free discussions between equals of a kind necessary to promote variation, but are dominated by the authority of male elders with the result that change is possible only within the conservative limits endorsed by those elders).

There is nothing surprising in this. As the son of a Manchester non-conformist liberal family, Spencer was well-schooled in classical liberalism. He was, as a natural historian by training and an ethnographer by vocation, equally well-schooled in Darwinian thought and its application to the fields of anthropology and archaeology. While never eschewing his earlier position his later explanations of the 'unimprovability' of Aborigines drew more on the terms of his racialisation of Aboriginality. This, as I argued in Chapter 4, inscribed backwardness in the body by interpreting the Aborigine as the product of a bloodline that had failed to respond to the dynamics of competition. The problem here, to recall Anderson's account, was that of being too far away in time to be susceptible to the influence of civilising programmes. However, this comprised less a shift from innatist conceptions of race than a historicisation of the basis on which innatist racial distinctions were drawn. In contrast to polygenetic accounts of innateness, Spencer and his contemporaries drew on the post-Darwinian traditions, discussed above, in which innateness had been historicised. Different races were the bearers of the different 'innatenesses' that they inherited as a consequence of the ways in which the dynamics of the relations between will, reason, habit and instinct had been played out in earlier generations.

It is in this respect that the doctrine of survivals – the keystone, according to George Stocking (1987), of late Victorian imperial anthropology – played such a crucial role in both the conception and administration of race in colonial contexts. Initially elaborated by Edward Burnett Tylor (1871) it organised what Patrick Wolfe calls the 'spatiotemporal triad' of imperial modernity, a triad consisting in '"our" (i.e. Europeans') savage past, "their" (i.e. colonised natives') ethnographic present, and "our" civilised present' (Wolfe 1999: 131). The aspect of this doctrine that is most relevant to my concerns here consists in the role it accorded rituals as part of a distinctive technique for deciphering the relations between past and present. Wolfe attributes this partly to anthropology's need for an object of analysis – rituals – and a technique of decipherment that would legitimate its claims to disciplinary autonomy by distinguishing its objects and methods from those of geology and philology, which provided the master discourses for interpreting the remote past by means of the material and/or textual forms it had left behind. He finds the model for Tylor's move, however, in Max Müller's 'disease of language' theory. Initially propounded in 1861, this argued that linguistic forms continued to circulate after their original meanings had been lost or had withered. Tylor latched on to the role Müller attributed to empty, mechanical repetition in accounting for the persistence of such withered forms of language use. For it suggested that rituals, too, might be construed as practices that had persisted through time in a similar 'withered' form and might therefore serve as extant carriers of their original meanings.

This, then, provided imperial anthropology with its distinctive disciplinary manoeuvre through which the analysis of current ritual practices could also serve as the means for reconstructing a prehistoric culture that still survived in the present. It was this disciplinary manoeuvre that presented the 'unimprovability' of Aborigines in a new light in their constitution as a site of both somatic and cultural 'flat-lining': that is, of persisting, like an electrical time-sequence measurement that shows no activity, constantly on the same level. This was not, however, because the persistence of rituals meant that the role of habit *per se* was too strong among 'primitive' peoples. The problem was rather that, in the case of 'the primitive', the dynamic set of relations posited by Ravaisson and later, in a more evolutionary framework, by Herbert Spencer through which responses to a changed environment are worked through from conscious action via habit into instinct so as to build up a pro- gressively accumulating set of instincts is blocked, locked in on itself, through the endless repetition of an original habit-into-instinct cycle. The consequence of this for Aborigines, paradoxically, was that they were depicted as having *too thin* a stock of instincts to be civilisable. Still on the cusp of the transition from nature to culture, Aboriginal conduct is interpreted as being guided by an original set of instincts – by, in Ravaisson's terms, a primitive rather than a secondary nature. To the degree that these have been repeated over the inter- vening millennia as survivals of an incomplete transition from nature to culture, so their power is increased by dint of the force of repetition with the consequence that they now exercise a more-or-less iron-like grip on conduct.

This logic is clearly discernible in Henry Pitt Rivers' account of the reasoning underlying his anthropological collection in which he adapts Tylor's account of survivals to interpret the tools and weapons of 'primitive' peoples as similarly survivals of earlier forms.[4] Drawing on Spencer's *Principles of Psychology* and Tylor's *Primitive Culture* as well as on John Lubbock's equally influential *Prehistoric Times* (1865), and presenting his argument as an evolutionary confirmation and extension of Locke's critique of innate ideas, Pitt Rivers con- strues the relations between habit and instinct in both animals and humans as being governed by essentially the same principles.[5] Just as the habits acquired by animals via either domestication or their reasoning on experience become instinctive and are passed on as such to their offspring, so similar processes are involved in the relations between the roles played by the 'intellectual mind' and the 'automaton mind' in regulating human conduct:

> We are conscious of an intellectual mind capable of reasoning upon unfamiliar occurrences, and of an automaton mind capable of acting intuitively in certain matters without effort of the will or consciousness.

> And we know that habits acquired by the exercise of conscious reason, by
> constant habit, become automatic, and then they no longer require the
> exercise of conscious reason to direct the actions, as they did at first.
>
> (Pitt Rivers 1875: 296)

The conclusion Pitt-Rivers draws from this is that 'every action which is now
performed by instinct, has at some former period in the history of the species
been the result of conscious experience' (298). This conception forms part of
a mechanism of development according to which the more that simple ideas
derived from experience are passed on into the automated forms of instinct,
the freer the person is to respond to new and more complex ideas. The key
hinge in this mechanism is habit, which Pitt Rivers interprets as a form of
conscious learning involving the intellectual mind but which then becomes
routinised via repetition. It is through habit that the lessons of experience are
passed on into instinct in accordance with an accumulative logic in which the
completion of one habit-to-instinct cycle frees up the space for another such
cycle, leading to an ever-growing set of instinctual responses constituting the
automated mind.

The colonial sting in the tail of this argument comes when Pitt Rivers
argues that 'the tendency to automatic action upon any given set of ideas will
be in proportion to the length of time during which the ancestors of the
individual have exercised their minds in those particular ideas' (299). This is
why lower animals, whose instincts have not been modified to the same degree
as those of higher animals, are more predisposed towards automatic forms of
action: they have practised the same set of automated responses for longer,
with a consequent increase in their hold on behaviour. The position of the
Aborigine is broadly similar. Poised forever on the cusp of the nature/culture
divide, the Aborigine never moves beyond simply imitating natural forms and
adapting these for certain purposes (Pitt Rivers accounts for the development
of Stone-Age tools in these terms), which are then performed repeatedly
across generations. The consequence is that 'in proportion to the length of
time during which this association of ideas continued to exist in the minds of
successive generations of the creatures which we may now begin to call men',
then so 'would be the tendency on the part of the offspring to continue to
select and use these particular forms, more or less instinctively – not, indeed,
with that unvarying instinct which in animals arises from the perfect adaptation
of their internal organism to the external condition, but with that modified
instinct which assumes the form of a *persistent conservatism*' (300). For the
savage and especially, as Pitt Rivers' paradigm of savagery, the Aborigine, the
problem is that the mechanism of habit has not worked with sufficient vigour
to build up an accumulated stock of 'modified instincts' but only a thin layer
of these, which, due their endless repetition over millennia, have acquired an
unusually binding grip on conduct. Pitt Rivers does not cite him, but Bagehot's
formulations point in the same direction. When he asks what the difference is
between prehistoric man and 'modern-day savages', Bagehot answers that the

former were 'savages without the fixed habits of savages' (Bagehot 1873: 113). In all other respects identical, prehistoric man

> differed in this from our present savages, that he had not had time to ingrain his nature so deeply with bad habits, and to impress bad beliefs so unalterably on his mind as they have. They have had age to fix the stain on themselves, but primitive man was younger and had no such time.
>
> (143–5)

As an armchair anthropologist, Pitt Rivers wrote at a distance from the immediacies of colonial rule, as did Tylor, Bagehot and Maudsley. Nonetheless, their formulations contributed to the organisation of the discursive ground that mediated the relations between the 'settler' and the Aboriginal populations in late-nineteenth- and twentieth-century Australia. These took distinctive forms governed by the logic of settler colonialism, which, since its primary object is possession of the land rather than the surplus to be derived from mixing indigenous labour with the land, aims at the elimination of the indigenous population. In the Australian case, Patrick Wolfe (1999) argues, this structure has taken three forms: that of frontier confrontation aimed at the annihilation of the colonial population; incarceration pending the inevitability of the Aborigines' extinction faced with competition from a superior race; and assimilation via managed programmes of epidermal and cultural integration with the white population. Beliefs in the 'unimprovability' of Aborigines figured prominently in the second and third stages where they operated in accordance with the imperatives of biopower, according to which the power to 'make live' by improving the health and conditions of life of the population is counterbalanced by the right to 'let die' by eliminating 'the biological threat to the improvement of the species or race' (Foucault 2003: 256). It is for this reason, Foucault suggests, that evolutionism played such a key role in nineteenth-century colonial practice:

> Whenever, in other words, there was a confrontation, a killing or the risk of death, the nineteenth century was quite literally obliged to think about them in the form of evolutionism. If you are functioning in the biopower mode, how can you justify the need to kill people, to kill populations, and to kill civilisations? By using the theme of evolutionism, by appealing to a racism.
>
> (257)

Evolutionary accounts of the mechanisms through which habit is translated into instinct provided a warrant for the exercise of biopower where, in the case of 'primitive' peoples, the regular functioning of these mechanisms had been blocked. For the form of repetition that this embodies generated what Pitt Rivers called the insuperable problem of sequence:

> Or two nations in very different stages of civilisation may be brought side by side, as is the case in many of our colonies, but there can be no

amalgamation between them. Nothing but the vices and imperfections of
the superior culture can coalesce with the inferior culture without break
of sequence.

(Pitt Rivers 1875: 308)

This is precisely the problem posed by survivals: they run against the tide of the
competitive struggles responsible for the 'survival of the fittest'. The survival,
Tylor argued, is an exception to the 'slow process of natural selection, ever
tending to thrust aside what is worthless, and to favour what is strong and
sound'; it is a 'stream of folly' that 'the savage' carries 'far into the culture of the
higher races' (Tylor 1867: 92). Distinct from the continuity of tradition or the
episodic logic of the revival, the survival is out of place in the present; it is, in
evolutionary terms, anachronistic. As the trace of a once-useful but now redun-
dant set of practices, the survival represents what Georges Didi-Huberman
usefully calls 'spectral time' (Didi-Huberman 2002: 62), a past in the present
that, having outlived its usefulness, does not properly belong there and, more
to the point, should not remain there.

For Tylor's purpose was not that of a sentimental folklorist proposing that
survivals should be preserved as curios. The doctrine of survivals is not a
'mere abstract truth, barren of all practical importance' (Tylor 1867: 93). To
the contrary, it is the means through which ethnography sifts out those ideas,
opinions and customs that have a role to play in enhancing progress and those
that are a fetter on the dynamics of social and cultural evolution. Tylor thus
concludes that 'the study of the lower races has a work to do in facilitating
the intellectual progress of the higher, by clearing the ground, and leaving the way
open for the induction of general laws and their correction by the systematic
observation of facts, to the results of which method alone we may fittingly
give the name of Science' (93). Indeed, he speculates that the continued existence
of savages might be accounted for, teleologically, as an act of providence that
has preserved them in order to serve this scientific purpose (Tylor 1867a: 314).
There is, however, only a short period of time available for this work to be
conducted. Reflecting that it 'may be the duty of civilised life and certainly its
effect, to put an end to savagery in the world', Tylor goes on to note that a
corollary of this is that our 'knowledge of savage life' has largely been
acquired 'in the process of improving them [savages] off the face of the earth'
(Tylor 1869: 23). Maudsley strikes a similar note when querying the rationality of attempts to enforce the habits of higher levels of civilisation on people
of a lower level. The very attempt to civilise savages, he suggests, constituted
a kind of ruse in which nature and culture conspire to translate the imperatives
of competition into their inevitable outcome:

However, as the thing is persistently and pertinaciously done by the
higher people moved by a holy impulse to confer the blessings of their
civilisation and religion, albeit at the cost of the destruction of the lower
peoples, we may conclude that the disintegration of the social structure

inevitably produced and the demoralisation of the people by the disorganisation of the cerebral reflexes constituting their mental fabric and serving their needs, are the ordained means by which nature degrades and finally eliminates the weaker races of men and promotes the survival and growth of the stronger races. And although the lower peoples may not feel happy to serve only as organic steps to build up a higher people, yet there is no help for it, they must suffer and die that the race may live and be strong.

(Maudsley 1902: 231–2)

It is often rightly objected that Foucauldian accounts of governmental rationalities pay insufficient attention to the more variable, muddied and muddled administrative arrangements that result from their translation into actual political programmes and policies (Clarke 2009). Maudsley's text was written in 1902, a year after the Federation of Australia, after which the earlier logics of settler colonialism progressively gave way to that of assimilation in the context of the development of a national governmental project and the associated formation of what Tim Rowse (1998) calls 'an Aboriginal domain' that aimed to integrate the Aboriginal population within the state. This was, however, no simple transmission with, after 1901, a significant variety of administrative arrangements continuing to order the relations between white and black Australia (Haebich 2000; McGregor 1997). The same was true of the discursive mediations of white/Aborigine relations that continued to draw on the mixed legacy of nineteenth-century polygenetic and evolutionary conceptions. However, to recall my earlier discussion of these questions in Chapter 4, two tendencies stand out. The first consists in the progressive marginalisation of those nineteenth-century missionary and philanthropic initiatives that, in some cases with government support, had aimed to civilise Aborigines, or to help them civilise themselves, by providing means for them to gain access to the resources of Christian and European civilisation. The second consisted in the exercise of new forms of biopower that aimed, through the strategy of assimilation, to breed out the race by separating 'half-castes' – now viewed, in a new light, as improvable on the grounds that their mixed bloodlines meant that they no longer posed an insuperable problem of sequence. 'Half-castes' were to be civilised in special stations designed for their improvement, and intermarriage between them would lead to the progressive dilution of the race across generations. In the meantime, 'full-bloods' were to be left to go their own way as decaying survivals.

Of course, there were many aspects to white accounts of Aboriginal backwardness: the fragility of their social and cultural forms; the morally and physically deteriorating consequences of their susceptibility to the 'vices of civilisation' and so on. It is, however, in the frequent reference to their inability to adapt, to their inertia, that the historical force of the habit–instinct nexus was evident in the new twist it gave to earlier discourses of 'unimprovability'. Tylor's view of primitive societies as anachronistic survivals did not, Didi-Huberman (2002) convincingly argues, amount to a total denial of any historicity whatsoever to such societies. Henrika Kuklick makes a similar point in relation to Baldwin Spencer, contending that he rebutted white settler views of Aborigines as a people without a history by – as we saw earlier – recording instances of conscious innovation and gradual improvement in numerous areas of Aboriginal life (Kuklick 2006: 562–5). However, I think that she misses the mark so far as the implications of Spencer's racialisation of Aboriginality are concerned. The historicisation of character developed across the human and natural sciences in the wake of Darwin's work made it possible to both recognise Aboriginal historical and cultural agency in the past and yet still place Aborigines on the other side of a historical divide from the white settler in the present. For the problem, as we have seen, was not that the Aborigine was innately incapable of either self-improvement or of being improved, but that he had *become* so. Although the result of a particular set of circumstances (the absence of competition), this incapacity was nonetheless interpreted as having become racially constitutive, inscribed within a separate bloodline, which meant that the capacity for innovation and volition that Aborigines had once shown could not vouchsafe the race a future. The continuing effects of this evolutionary habit–instinct nexus are evident in the formulations of the anthropologist A.P. Elkin. In 1932 – expressing a view he was later to revise – he attributed the failure of Aborigines to adapt to the requirements of a more advanced culture to their inherited racial constitution:

> We are almost forced to realise the possibility that the aboriginal race may have been so completely adapted biologically as well as mentally to its own cultural environment that it cannot adapt itself to a culture of a different type, or, in other words, that it lacks the 'ethnic capacity' to become civilised.
>
> (Elkin, cited in McGregor 1997: 199)

We can hear here the legacy of a very peculiar, and deadly, set of relations between liberal modernity and imperial Britain that reached far beyond Britain's shores and outlasted the imperial phase proper. Its logic as a form of governmentality depended on a particular ordering of the relations between habit, will and instinct that, instead of functioning as a coercive mechanism at the heart of liberal forms of self-rule, laid out the relations between races in ways that organised the exercise of a distinctive form of biopower. The opposition that habit was caught up in here was not one in which Aborigines were to be subjected to the mechanisms of drill and discipline rather than be treated as persons capable of self-governance through the exercise of will. This set of options was no longer on the agenda. The issues were rather posed in terms of survival with the habit–instinct nexus and its role in relation to the problem of sequence, defined in terms of bloodlines, guiding where the dividing line should be drawn between where the powers to 'make live' or 'let die' should be exercised.

9 Habitus/habit
Freedom/history

It is, on the face of it, an abrupt transition from the role played by the nineteenth-century concept of habit in mediating encounters across Australia's colonial frontier to consider the relationships between habit and the account I offered in Chapter 8 of aesthetics as a form of 'guided freedom'. And all the more so given that I shall pursue these questions by examining aspects of Bourdieu's account of the relations between aesthetics, habit and habitus. I do so with a view to showing how the ways in which Bourdieu distinguishes habitus from habit opens up, in the habitus, a space for a certain kind of freedom, but one that needs to be guided if it is to be led to its proper end. I shall also argue that Bourdieu's account of the uneven distribution of this capacity for freedom across classes is a sociologised variant of the ways in which philosophical aesthetics has, from the civic humanists onward, connected questions of taste to questions of governance. Equally, though, it is in his aspiration to broaden access to the capacity for a certain kind of freedom that has come to be coded into the aesthetic that we witness the respects in which, in spite of appearances to the contrary, Bourdieu's work is subtended by a Kantian narrative of human self-realisation, albeit one that is given a distinctive sociological twist.

I shall, in pursuing these lines of argument, aim to shift the register in which Bourdieu's key concepts are usually debated in order to bring to light those qualities that, although usually discounted as philosophical distractions from the methodological core of his empirical sociology, run throughout his work and have a crucial bearing on the forms of intellectual authority he sought to constitute and to exercise. My approach to his concept of habitus will thus leave to one side those concerns that interpret it, as Bourdieu did, as an alternative to the polarities of structure and agency, to focus instead on its less-frequently noted role in relation to the mechanisms of inheritance through which the accumulated history of the past is transmitted to the present and, in the process, creatively modified. This will entail an assessment of the points of entry into different architectures of personhood that Bourdieu's concept of habitus opens up to the collective intellectual. Similarly, rather than following the usual emphasis that interprets Bourdieu's critique of Kant's account of the disinterestedness of the aesthetic as a mask for particular social interests,

I shall highlight the implications of those aspects of his work that reflected a continuing, but reformatory, commitment to some of the key underlying principles of Kant's aesthetic. My focus will be on Bourdieu's attempts to historicise and sociologise these principles and, in the process, to construct, in his conception of the collective intellectual, a form of authority that blends the competencies of the philosopher-aesthetician and the sociologist. Rancière's comparison of Bourdieu to the sociologist-king – discussed in Chapter 7 – is, in this respect, misleading: the spaces between habit and habitus, between given tastes and those which are 'historically universal', which Bourdieu presents us with are ones within which collective intellectuals are to act as freedom's guides. It is Kant, not Wolff, who provides the relevant point of historical and philosophical reference here.

To consider Bourdieu's work in the light of these registers is, in sum, to engage with him as a liberal social and political thinker whose work belongs to that historical configuration of the relations between making culture, organising freedom and changing society that I discussed in Chapter 2.[1] It is to suggest, moreover, that he operated within these relations in a Kantian mode. Bourdieu invites us to see his work in this light often enough. In elaborating his understanding of the principles of a reflexive sociology, he stresses that its role is to identify 'true sites of freedom' and thus to build 'small-scale, modest, practical morals in keeping with the scope of human freedom' (Bourdieu and Wacquant 1992: 199). This scope, Bourdieu goes on to say, is not very large and can, moreover, only be won via knowledge of the determining force of necessity and thus of the limits to freedom and how, as a historical conquest, these might be overcome. It is, he says, 'through knowledge of determinations that only science can uncover that a form of freedom which is the condition and correlate of an ethic is possible' (198). When agreeing with Loïc Wacquant that such freedom is not the unconstrained freedom of a Cartesian cogito but 'a freedom collectively conquered through the historically dated and situated construction of a space of regulated discussion and critique' (190) – precisely, as we have seen, Kant's interpretation of Enlightenment – what is most compelling about Bourdieu's reply is how he immediately connects this question of freedom to that of the production of a 'universal subject' as 'a historical achievement that is never completed once and for all' (190). How is this historical narrative of freedom organised? What role does the concept of habitus play within it? What are its effects? These are the central questions I address in this chapter.

I begin, though, by looking at what Bourdieu has to say about archaic habitus. However much his concern may have been to extend freedom's remit as far as possible and without restrictions, Bourdieu's conception of the relations between history, habit and habitus also resonates with those that informed the nineteenth-century doctrine of survivals. A consideration of these questions will serve to show how, at times, Bourdieu's formulations are awkwardly installed within the problem-space that was produced by the late-nineteenth-century rendezvous between liberal and evolutionary thought that was mediated through the concept of habit. This is not, though, particularly surprising if we recall Bourdieu's training as an ethnologist schooled in the Durkheim–Mauss tradition that, as we have seen, was itself awkwardly installed in the very same problem space.

Archaic habitus

'Social agents, in archaic societies as well as in ours', Bourdieu says, 'are not automata regulated like clocks, in accordance with laws which they do not understand'. Rather, in their games, matrimonial exchanges and ritual practices, 'they put into action the incorporated principles of a generative habitus' (Bourdieu 1990: 9). Yet when, in another context, he refers to the members of archaic societies, he draws a line that inscribes 'them' and 'us' in a different relationship to time. It is a line that depends on the doctrine of survivals, and the occasion for drawing it is given by Bourdieu's concern to identify a point of leverage within the makeup of the modern person through which we might be freed from the inheritance of masculine domination by subjecting that inheritance to a historical anamnesis. This is the unlikely role in which he casts the Berbers of the Kabylia in interpreting them as the bearers of

> preserved structures which, protected in particular by the relatively unaltered practical coherence of behaviours and discourses partially abstracted from time by ritual stereotyping, represent a paradigmatic form of the 'phallonarcissistic' vision and the androcentric cosmology which are common to all Mediterranean societies and which survive even today, but in a partial and, as it were, exploded state, in our own cognitive structures and social structures.
>
> (Bourdieu 2001: 6)

There are two aspects to the temporality Bourdieu posits here. According to the first, all 'Mediterranean societies' represent, in matters of sexuality, archaic codes that have been preserved owing to the undue force of ritual stereotyping that have elsewhere in Europe, at least on the surface of things, fallen into disuse. Yet, in the second aspect of this temporality, the members of archaic societies also represent a universal aspect of 'our' (whose?) inheritance from our deep, archaeological past, one in which that past still operates within us as a level of unconscious determination from which we can only be freed by bringing it to consciousness. This is the task that Bourdieu calls on ethnology to perform. In its description of remote and distant social worlds, ethnology acts as

> a kind of 'detector' of the infinitesimal traces and scattered fragments of the androcentric worldview and, consequently, as the instrument of an archaeological history of the unconscious which, having no doubt been originally constructed in a very ancient and very archaic state of our societies, inhabits each of us, whether man or woman.
>
> (54)

It is by bringing this past that has survived into the present to light that 'ethnology constructs the potentially liberatory axiomatics' (55) needed to cast off the inheritance of masculine domination.

Bourdieu once said that he never tired of quoting Durkheim to the effect that 'the unconscious is history' (Bourdieu 2004: 96). He does so, for example, in *Outline of a Theory of Practice* where, in elaborating his conception of the habitus as 'history turned into nature' (that is, as history which is denied as such in its taken-for-grantedness), he argues that the unconscious is constituted in its forgetting of the history that has gone into the making of the second nature of the habitus. He cites, in illustration of this contention, Durkheim's assessment that although

> it is yesterday's man who inevitably predominates in us, since the present amounts to little compared with the long past in the course of which we were formed ... we do not sense this man of the past, because he is inveterate in us; he makes up the unconscious part of ourselves.
>
> (Durkheim, cited in Bourdieu 1977: 79)

It is, then, worth recalling that Durkheim drew substantially on Spencer and Gillen's fieldwork and on Tylor's doctrine of survivals for his account of the elementary forms of religious life (Durkheim 1968). He also endorsed Théodule Ribot's construal of neuro-physiological mechanisms, similar to those of Bagehot's 'connective tissue of civilisation', as being chiefly responsible for the mechanisms through which the past is preserved and transmitted into the present (Ribot 1997). These mechanisms, Durkheim argues, work via the 'organico-physical conditions' through which the 'native talents that are transmitted to us by our ancestors ... chain us, then, to our race ... and shackle the liberty of our movements' (Durkheim 1964: 304). Their influence on conduct is preponderant unless they are countered by the opposing force of individuality, an attribute which Durkheim denies the primitive.

We are all affected in ways that we are not aware of by the legacies of the intellectual traditions we work with and against in shaping our own thought. Bourdieu is no exception. It is clear, from the re-evaluations of his relations to colonialism that are currently underway (Goodman and Silverstein 2009), that Bourdieu occasionally recapitulated that denial of the coevalness of distant others that Johannes Fabian (1983) has argued characterises the colonial structure of anthropological discourse. George Steinmetz (2011) has noted this aspect of Bourdieu's work as, indeed, have I in commenting on his construal of the forms of masculine domination he identified in Algerian peasant society in the 1950s as archaic precursors of those evident in the 1930s Bloomsbury circle (Bennett 2005). This same denial of coevalness informs his account of the different ways in which the archaic habitus and the habitus of modern social agents are marked by time. The archaic habitus is not innately different from the latter; it becomes so. Habitus, in effect, becomes habit as conduct petrifies through the enforced repetition of ritual. The architecture of the archaic person thus lacks that tension between inheritance and the generative capacity of the habitus within which, however limited it might be, a capacity for freedom is produced. And if ethnology is indeed to contribute to the work of historical anamnesis through which freedom is to be produced as a historical outcome rather than treated as a given, then it is, at least in the first instance, 'our' freedom, not 'theirs', that is at issue in studying the habitus of those who lived, or live, in archaic societies, past or present.

There is, then, a lack of symmetry between the structures of archaic and modern habitus so far as their relations to historical time are concerned. It is only the latter that are potentially capable of reflecting back on the history of their own determination to produce a space of undetermined action in which the subject is (relatively) free to produce its own future. Appealing to the Thomist interpretation of Aristotle, Bourdieu interprets the habitus – which he distinguishes from habit as, in his (largely Cartesian) interpretation, mere mechanical repetition – as the locus where this kind of (relatively) free reshaping of the historical forces that have shaped the person takes place.[2] This conception of the habitus as a site for the mediation of the relations between an accumulated history and its situationally motivated activation in the present is given, in one of Bourdieu's more influential formulations, in the form of a distinction between 'two states of history'. There is, first, the objective history that is accumulated across the course of time in 'things, machines, buildings, monuments, books, customs, rights, etc.' and, second, 'history in its embodied state, having become habitus', where habitus is interpreted as the 'product of a historical acquisition which permits the appropriation of that acquisition' (Bourdieu 1980: 6). Returning to these formulations in *Pascalian Meditations*, Bourdieu interprets the habitus as the mechanism that equips the inheritor to inherit his/her inheritance. 'The inherited inheritor, appropriated by his heritage', he says, 'does not need to *want*, in the sense of deliberating, choosing and consciously deciding, in order to do what is appropriate, what corresponds to the interests of the heritage, its conservation and its increase' (Bourdieu 2000: 151). Yet this is not a matter of pure automaticity either: the adoption of one's inheritance proceeds through the mechanism of the habitus as, in Bourdieu's famous formulation, a 'structured and structuring structure' that partially remoulds the conditions which condition it (Bourdieu 1984: 171).

Habitus, however, are not uniform in the manner of their formation or functioning. Some allow a greater capacity for reflexively modifying the force of the determinations that condition them than do others. The extent to which this is so depends on the distance that particular habitus open up in relation to habit and the opportunity this provides for 'authorities of freedom' to guide and direct the capacity for freedom that this distance produces. This is the space that Bourdieu seeks to occupy on behalf of, depending on the context, the sociologist or the collective intellectual. He thus concludes his 1980 essay on the relations between reified and incorporated history with an account of how the sociologist, in accounting for the capacity of the habitus to reshape itself, makes possible a certain 'mastery of the self' for both individuals and for groups to the extent that 'the scientific knowledge of necessity contains the possibility of an action aimed at neutralising it, and thus a *possible* liberty' (14). In *Pascalian Meditations* and *Masculine Domination*, the same possibility is opened up by the practice of historical anamnesis. It is through this that the collective intellectual brings to light the hidden forces that have shaped the path of social and cultural development so as to open up a space for a practice of freedom that will place those forces under the direction of a self-reflexive collective historical agent. This collective intellectual is another 'authority of freedom' deciphering the conditions under which freedom can emerge and be exercised, by whom, and, correlatively, who must be denied it until the historical conditions for its universalisation have been produced. The same is true of the postscript to *The Rules of Art* where Bourdieu urges intellectuals to accept responsibility for a project of historical anamnesis as part of a politics of freedom through which writers, artists and intellectuals will seek, first, to recover the history of the struggle for a collective universal that is implicated in the struggle for artistic and intellectual autonomy, and, second, to defend that autonomy against the state and market while simultaneously seeking to extend its social reach. This is, of course, a formulation of the politics of freedom that is unimaginable without the role that Kant envisaged for aesthetic judgement as itself a practice of freedom whose exercise involved the historical projection of a *sensus communis*.

These differences between habitus, however, do not just concern the relations between archaic and modern habitus. They also concern distinctions beween modern habitus. This is true of Bourdieu's classic account, in *Distinction*, of different class habitus. This is usually assessed as a sociological account of the relations between three class habitus: the ethos of disinterestedness of the bourgeoisie, the culture of goodwill of the petit-bourgeoisie, and the working-class culture of the necessary. Whatever its strengths and weaknesses in this regard,[3] *Distinction* is also informed, albeit at a more subterranean level – that of its aesthetic unconscious, if you like – by the legacy of the relations between aesthetics and liberal political reasoning that is integral to the structure of Bourdieu's thought. My interest, from this perspective, concerns the manner in which Bourdieu distributes differently weighted capacities for being governed by freedom and by necessity across different class habitus, and the ways in which, in doing so, he remains in the slipstream of aesthetics as a discourse that qualifies some for political participation at the price of disqualifying others.

Aesthetic and political disqualification

Let's first look in a little more detail at Bourdieu's assessment of the degree to which habitus are unified. He places considerable stress on the unity of the habitus in his earlier and more programmatic elaborations of the concept. In *Practical Reason*, Bourdieu defines the habitus as a 'generative and unifying principle which retranslates the intrinsic and relational characteristics of a position into a unitary lifestyle, that is, a unitary set of choices of persons, goods, practices' (Bourdieu 1998: 8). And in *Distinction*, writing about the aesthetic disposition as an aspect of the system of dispositions comprising a class habitus, he could not be more emphatic:

> Being the product of the conditionings associated with a particular class of conditions of existence, it unites all those who are the product of similar conditions while distinguishing them from all others. And it distinguishes in an essential way, since taste is the basis of all that one has – people and things – and all that one is for others, whereby one classifies oneself and is classified by others.
>
> (Bourdieu 1984: 56)

It is true that there are places where Bourdieu opts for more elastic inter-pretations. In *Outline of a Theory of Practice* Bourdieu is more cautious in stressing the probabilistic expectation that members of the same class will share the same experiences, stressing that not all the members of a class – or even two of them – will share the same experiences acquired in the same order (1977: 85).[4] Similarly, the regularity he prescribes for the habitus in *In Other Words* is one in which *'the habitus goes hand in glove with a vagueness and indeterminacy'*; it has 'a generative spontaneity which asserts itself in an improvised confrontation with ever-renewed situations', obeying 'a *practical logic*, that of vagueness, of the more-or-less, which defines one's ordinary relation to the world' (Bourdieu 1990: 77–8). It is also true that he usually allows for a tension, in the case of artists and intellectuals, between their class habitus and the habitus associated with their distinctive position in a specific artistic or intellectual field – a position he proposes in his account of the scien-tific habitus, for example (Bourdieu 2004). And in *Pascalian Meditations* finally, the habitus emerges in an utterly transformed form as full of 'mismatches, discordances and misfinrings' such that those who occupy contradictory social positions often have 'destablised habitus, torn by contradiction and internal division, generating suffering' (Bourdieu 2000: 160).

However, these later qualifications and revisions do not alter the fact that, in *Distinction,* the generative schemas of the habitus apply across different fields of consumption through a simple mechanism of transference. 'The practices of the same agent', he writes, 'and, more generally, the practices of all agents of the same class, owe the stylistic affinity which makes each of them a metaphor of any of the others to the fact that they are the product of transfers of the same schemes of action from one field to another' (Bourdieu 1984: 173). His account of the homology between positions in the space of lifestyles means that the principles underlying an individual's or group's tastes in the literary field are thus also held to apply to that person's or group's tastes in all other fields – the musical and artistic fields, for example. It is this mechanism, Bourdieu argues in *Distinction*, that is made manifest in the systemacity that is to be found across all aspects of an individual's tastes:

It is be found in all the properties – and property – with which individuals and groups surround themselves, houses, furniture, paintings, books, cars, spirits, cigarettes, perfume, clothes, and in the practices in which they manifest their distinction, sports, games, entertainments, only because it is in the synthetic unity of the habitus, the unifying, generative principle of all practices. Taste, the propensity and capacity to appropriate (materially or symbolically) a given class of classified, classifying objects or practices, is the generative formula of life-style, a unitary set of distinctive preferences which express the same expressive intention in the specific logic of each of the symbolic sub-spaces, furniture, clothing, language or bodily hexis.

(Bourdieu 1984: 173)

The unity Bourdieu attributes to the habitus, however, is a principle that Bourdieu applies far more rigorously in relation to the working class than to other classes. In his discussion of the relations between habitus and the space of lifestyles, Bourdieu identifies two main factors structuring the habitus: first, the intrinsic properties that derive from the conditions of existence of the class concerned, and second the relational properties that each class derives from its relations to other classes within the system of differences which constitutes them such that the whole system of such relations is inscribed within each habitus. However, the balance between these two factors turns out to be unevenly distributed across different classes. The position that is available to the working class within this scheme is one of functioning as a fixed point of reference in relation to which other positions differentiate themselves. It is not itself a site of differentiating activity except purely negatively as, in the process of rejecting the dominant culture, the working class is thrown back on its own pure class conditioning. They are not players in the field in which the game of distinction takes place; rather, they provide the setting against which that game is played, but only by other players. While, as we have seen in Chapter 7, there are good reasons to be cautious regarding the oracular hermeneutic that informs his ventriloquism of the oppressed, Rancière is right in his assessment of this aspect of *Distinction*:

As for the poor, they do not play. Indeed, their habitus discloses to them only the semblance of a game where the anticipated future is not what is possible but simply the impossible: 'a social environment' with 'its "closed doors," "dead ends" and "limited prospects"' where 'the "art of assessing likelihoods"' cannot euphemise the virtue of necessity. Only those who are chosen have the possibility of choosing.

(Rancière 2004: 183)

The working-class habitus is, as a consequence, far more singularly unified and fixed than are other class habitus.[5] For the working class, the dire weight of necessity permits not the slightest chink of differentiation. Whenever Bourdieu speaks of a working-class aesthetic, 'aesthetic' is placed in quotation marks as, strictly speaking, a contradiction in terms since it lacks any element of free or conscious choice that is required for the exercise of aesthetic judgement. For example:

> ... nothing is more alien to working-class women than the typically bourgeois idea of making each object in the home the occasion for an aesthetic choice ... or of involving specifically aesthetic criteria in the choice of a saucepan or cupboards.

> ... rooms socially designated for 'decoration', the sitting room, dining room or living room ... are decorated in accordance with established conventions, with knick-knacks on the mantelpiece, a forest scene over the sideboard, flowers on the table, without any of these obligatory choices implying decisions or a search for effects.

> Perhaps the most ruthless call to order ... stems from the closure effect of the homogeneity effect of the directly experienced social world. There is no other possible language, no other life-style, no other form of kinship relation; the universe of possibles is closed.

> Those who believe in the existence of a 'popular culture' ... must expect to find ... only the scattered fragments of an old erudite culture (such as folk medicine), selected and reinterpreted in terms of the fundamental principles of the class habitus and integrated into the unitary world view it engenders.
> (Bourdieu 1984: 379, 381 and 395)

In the chapter immediately following his discussion of the working-class choice of the necessary, Bourdieu discusses the working class's relation to politics. Taking issue with those who, throughout his career, he variously described as populists or class racists who practised an inverted ethnocentrism by crediting the common people with an innate knowledge of politics (Bourdieu 1984: 374; Bourdieu 2000: 76), Bourdieu's concern is with the processes through which the working classes are both politically disqualified, and disqualify themselves, because they lack access to the appropriate means of forming political opinions and judgements. He sets this topic up by referring to Marx's and Engels' projection of communist society as one in which it is only the generalisation of freedom from necessity across all classes that, by making participation in aesthetic practice and judgement possible for all, also allows 'everyone sufficient free time to take part in the general affairs of society – *theoretical as well as practical*' (Marx and Engels, cited in Bourdieu 1984: 397). This reworks, as a historical project, the ways in which, in civic humanist aesthetics, the connections between aesthetic judgement and the political qualification of the gentry on the one hand had been defined against the political disqualification of the artisan classes on the other. The difficulty is that, pending the equalisation in the distribution of free time that Marx and Engels anticipate, the working class is disqualified from political agency in their work too. Or rather, as Rancière argues, the proletariat is called on to fulfil a demand for political agency (the revolution) that, in its given forms, the proletariat cannot discharge until it has acquired the capacity to do so. This, however, is not a capacity the proletariat can develop itself; rather, it must receive it in the form of a gift by being impregnated by philosophy and thus, in practical terms, being subordinated to the direction of intellectuals.[6] In the meantime, in its actual empirical forms, in an argument that was to be most fully developed in Lukács's account of the difference between actual and imputed class consciousness, the proletariat is disqualified from political action unless guided from without by the philosopher.[7]

My purpose is not to side with Rancière's contrived production of the 'direct voice' of the people here.[8] Rather it is to note how, in Bourdieu's account, the working classes, while possessing formal political rights, remain subject to informal political disqualification owing to their exclusion from the means of forming political opinions and judgement. He interprets this as an effect of their exclusion from the education system and the consequent limitation of their horizons to the uniformly flat and singular structure of a habitus shaped exclusively by the culture of the necessary that is the class's defining condition. Bourdieu's concern, of course, is to identify these informal mechanisms of political disqualification as forms of symbolic violence that need to be counteracted by the equalisation of educational opportunities. This provides the only means of universalising the conditions of access to the universal that he sees as a precondition for full and effective citizenship. The difficulties concern what is to happen in the meantime.

Bourdieu throws some light on this in his essay on what makes a class where he outlines the mechanisms that are needed to translate what he calls probable classes into actual ones. Probable classes are defined by their relations to one another within the relationally constituted space of positions, and are occupied by agents who are 'subject to similar conditions of existence and conditioning factors and, as a result, are endowed with similar dispositions which prompt them to develop similar practices' (Bourdieu 1987: 6). On the other hand, whether or not a probable class becomes an actual class such that the way in which its projects potentially divide the space of positions prevails over other potential divisions (those of gender or ethnicity, for example) depends on whether or not a political process of class making (he cites E. P. Thompson's work favourably here) is able to produce the class as 'a *well-founded artefact*' in the same sense that Durkheim spoke of religion as a 'well-founded illusion' (8–9). But this can only be the result of a process of delegation that is simultaneously one of dispossession:

> So in order to give a brief answer to the question posed, we will say that a 'class,' social, sexual, ethnic, or otherwise, exists when there are agents capable of imposing themselves, as authorised to speak and act officially in its place and in its name, upon those who, by recognising themselves in these plenipotentiaries, by recognising them as endowed with full power to speak and act in their name, recognise themselves as members of the

class, and in doing so, confer upon it the only form of existence a group can possess.

(15)

A similar kind of delegation is involved in Bourdieu's view that intellectuals have a responsibility to defend and promote the interests of the universal, which they are able to distil from the past struggles of artists and intellectuals and to accumulate into the present via a corporate act of anamnesia, while the working-classes' capacity for political agency is mired by the force of necessity. Brian Singer makes the point forcefully:

> The universal cannot be universal unless the excluded are included, but their voice is not the voice of the universal and does not, apparently, contribute to its expression. Thus they must be represented by what in another book he calls 'the corporation of the universal', that is, by intellectuals. But how is such a logic of 'representational substitution' (which has always been inherent to the s*kholè*) to overcome the divisions it decries when the division between representatives and represented is so great that the represented appear totally incapable of representing themselves (because they lack the dispositions and aptitudes, not to mention the socially validated competencies). The only solution appears to lie in a pedagogical utopia wherein the logic of substitution is gradually replaced by a logic of representational absorption such that everyone is able to speak the language of the universal.
>
> (Singer 1999: 294)

It is, for Bourdieu, aesthetics that provides this language of the universal. The question I now go on to consider is how he hitches this to his interpretation of freedom as a task to be accomplished.

Accomplishing freedom

In the chapter on aesthetics in his *Manual of Ethnography*, Marcel Mauss accepts the Kantian distinction between the useful and the beautiful in his definition of the aesthetic as 'a pleasure that is sensory but disinterested' (Mauss 2007: 68). However, contrary to Kant, he argues that 'the significance of the aesthetic in all the societies that have preceded us is great' (68). This reflects an elastic definition of the aesthetic as a multi-stranded aptitude that might take many forms. It might take the form of the disinterested contemplation of an object abstracted from need. It might take the form of enjoyment for its own sake: the 'enthusiasm and catharsis' of Zuni festivals leading to purging and vomiting is the case he cites. But there can, he says, 'be no beauty without sensory pleasure', citing the example of the Australian *corroboree* as a ritual which '"unravels the belly"' (69). While it is never absent, finally, Mauss advises that the sense of beauty is 'not found in all classes of the society in the same form, and it applies to different objects' (69).

Bourdieu endorsed Mauss' perception that the sense of beauty is differentially distributed across classes with regard to both its forms and objects. He did not, though, share his view that a capacity for aesthetic appreciation is *in fact* universal or that it might take such a diversity of forms as Mauss suggested. From *The Love of Art* through *Distinction* to his later work, Bourdieu, while criticising many aspects of Kant's aesthetic, did not question its central claims: first, that an adequate appreciation of aesthetic form depends on distanced and reflective forms of appropriation that proceed independently of concepts;[9] and second, that these are *not* given as universal but are *to be made* universal. In *The Love of Art*, for example, Bourdieu stated that his intention was not to 'to refute Kant's phrase that "the beautiful is that which pleases without a concept"', but rather 'to define the social conditions which make possible both this experience and the people for whom it is possible (art lovers or "people of taste") and thence to determine the limits within which it can exist' (Bourdieu and Darbel 1991: 109). But these limits were barriers to be overcome through the education system that, in accordance with the traditions of French republicanism, Bourdieu repeatedly called on to distribute to all classes the aptitudes required for an adequate appreciation of, and engagement with, those forms of literary and artistic culture that have been vested with a universal value by the institutions of cultural legitimation. We find a similar perspective in *Practical Reason*:

> I am ready to concede that Kant's aesthetics is true, but only as a phenomenology of the aesthetic experiences of all those people who are the products of *skholè*. That is to say that the experience of the beautiful of which Kant offers us a rigorous description has definite economic and social conditions of possibility that are ignored by Kant, and that the anthropological possibility of which Kant sketches an analysis could become *truly universal* only if those economic and social conditions were universally distributed.
>
> (Bourdieu 1998: 135)

The '*datum* from which sociological reflection starts', he continues, 'is not the universal capacity to grasp the beautiful, but rather the incomprehension, the indifference of some of the agents who are deprived of the adequate categories of aesthetic perception and appreciation' (135). The uneven distribution of this capacity applies 'within our own societies, across social classes, or ethnic groups' and 'across civilisations, from the Trobriand Islands to the United States of today' (137). The conclusion Bourdieu draws from this is that for intellectuals or the state to support the development of a popular, working-class, black or (by implication) Trobriand Islander culture or aesthetic would be misleading on two counts. First, since such tastes are not convertible into tradable forms of cultural capital, such populist 'rehabilitations' of subordinate groups succeed only in '*pushing them further down* ... by converting deprivation and hardship into an elective choice' (137). Second, it detracts from the political demand that the full accomplishment of human potentialities that is represented by the Kantian conception of the disinterested appreciation of beauty requires a commitment to '*working to universalise the conditions of access to universality*' (137).

This is, however, the very essence of Kant's aesthetics so far as its translation into distinctive politico-aesthetic programmes is concerned. The conditions Kantianism generates for such programmes are that (i) the universality of the aesthetic must be projected as a goal to be accomplished, (ii) there must be agents of transition that can distil this universality in advance of its accomplishment and so serve as a conduit to it, (iii) the failure to judge competently must be accounted for in terms of a deficit in the makeup of the subject, and (iv) that such deficits can therefore only be overcome though the actions of some other agent.[10] That deficit may be, as in Bourdieu's case, benignly construed in the sense that it is accounted for in terms of social processes rather than as a result of the inherent limitations of innate dispositions. Nonetheless, the deficit that he attributes to the working class operates in similar terms to the deficits that earlier versions of Kantian aesthetics have attributed to other social groups where these, too, have been accounted for in terms of the constraints imposed by necessity.

The key issues here concern Bourdieu's conception of the relations between the education system, the role of intellectuals and the state. There is a consistency to his views on these matters from *The Love of Art*, through *Outline of a Theory of Practice*, *Distinction*, *The Rules of Art*, *Practical Reason* and *Pascalian Meditations*. They have three main components. The first concerns his expectation that the education system should function as the primary vehicle for universalising the conditions of access to the universal represented by canonised works. Second, the universality that is embedded in such works is not simply given but is rather the manifestation of a complex historicity through which those works transcend the particular conditions of their making not because of an inherent transcendentalism but by virtue of their relations to the historical dynamics underlying the development of the literary and artistic fields. Their 'historical universality' – to use Bourdieu's term – is a value bestowed on them by the collective intellectual whose autonomy from the economic and political fields that is produced by the historical dynamics of the literary and artistic fields vouchsafes for them a position of putative universality. It is by virtue of representing a universality that is in the processes of being made that such intellectuals are able, through a practice of historical anamnesis, to sift the cultural heritage of humanity so as to separate the genuinely universal from the irremediable particularity of other practices.[11]

How, though, are these universal values, once deciphered, to be disseminated so as to be made available to all and thus become universal in actuality? This third aspect of Bourdieu's account depends on his interpretation of the relations between collective intellectuals and the class to which, objectively, they belong but from which, through their practice, they distance themselves: that is, the dominated fraction of the dominant class. The programme Bourdieu envisages entails a reformation of the mode of appropriating culture that he attributes to this class that would sever the connections between the principles of 'pure taste', the schooling system and the occupational class structure by producing and circulating modes of appropriating culture that detach it from such class-specific *parti pris*. Since the dominated fraction of the dominant class uses its influence to reproduce the connections between culture, education and class on which its preferment depends, Bourdieu calls on the state to initiate and drive this process of reform. It is to do so through the agendas it sets for the institutions of cultural legitimation and the schooling system, and through the manner in which it acts on the relations between them. The state Bourdieu has in mind, however, is not that constituted by its given empirical forms – not the state as a site of complex political and bureaucratic processes, agents and calculations – but a state that has been subordinated to the direction of the collective intellectual. This, moreover, can be accomplished only provided that such intellectuals stand outside of the state so that, uncontaminated by its machinations and compromises, they can represent the universal over and against it.

Jeremy Ahearne has usefully highlighted the significance of those traditions of French intellectual life in which intellectuals have sought to exercise a form of 'counter-sovereignty' by claiming an authority to oppose short-term demands of political expediency in the name of longer-term demands associated specifically with the cultural realm.[12] The specific form of this position that Bourdieu enunciates replicates the deep structure of Kant's aesthetic in opening it up as a sphere of action that stands outside the state. Let me recall the four conditions I identified earlier that are required to translate Kant's analytic of the beautiful into distinctive politico-aesthetic programmes, and outline the manner in which Bourdieu meets each of these.

(i)   *That the universality of the aesthetic must be projected as a goal to be accomplished*: this, for Bourdieu, is a matter of the historical dynamics of the relations between the literary and artistic fields in generating a set of universal cultural values which have then to prevail over the forms of power associated with the economic and political fields if their universality is to become universally available.

(ii)   *That there must be agents of transition that can distil this universality in advance of its accomplishment and so serve as a conduit to it*: this is the role of the collective intellectual whose recall and selective sifting of the past identifies the tendential universality that is the product of past history and also a pointer to the path that history must pursue to make that universality actual.

(iii)   *That the failure to judge competently must be accounted for in terms of a deficit in the makeup of the subject*: this is the result of the class-specific partialities and limitations associated with different habitus.

(iv)   *That such deficits can therefore only be overcome though the actions of some other agent*: the pincer movement in which Bourdieu catches the working class means that it is unable to overcome the constraining effects of necessity though its own efforts. Such overcoming can only come from without in the form of a gift. The donor here is the collective intellectual albeit that the completion of this gift transaction is dependent on a series of intervening agencies – the school, the education system – if the universal that the collective intellectual identifies is to be relayed to the population at large.

Bourdieu begins his chapter on the working-class choice of the necessary by pitching his account of this habitus against two forms of *ouvriérisme*: against aesthetic forms of *ouvriérisme*, since the choice of the necessary comprises a modality which is 'not that of intellectual or artistic revolts' (Bourdieu 1984: 372), and against political forms of *ouvriérisme*. While arguing that the constraint that necessity – the 'inescapable deprivation of necessary goods' (372) – imposes on the working class is 'in no way incompatible with a revolutionary intention' (372), Bourdieu takes issue with the '*narodniki* of all times and all lands' for assuming that a working-class mentality can be deduced from empathetic identification with the position of the working class in the relations of production. This fails to take account, he argues, of the role of the habitus in providing the schemes of perception and action that mediate the class' relations to its position within those relations of production. And in calling this into question, Bourdieu's account of the habitus severs the connections that, in Marxist thought, cast the working class – once impregnated by philosophy – as the leading agents of a process of universalisation inscribed in the dynamics of the real.

There is, in Bourdieu's account, no movement of 'real history' that will lead the working class and the rest of society beyond the limiting effects of necessity and the schismatic effects of the division of labour. This is replaced by a narrative that works in terms of the relations between fields, the position of the collective intellectual within these, and the relations between such intellectuals, the state and the education system, through which the working class – and, indeed, other classes – will eventually receive the universal as an indirect gift from collective intellectuals. It is a narrative that depends for its deliverance on the role that such intellectuals play as mediators in the relations between habitus and habit, freedom and necessity. My purpose in making this point, though, is not to argue for a reversion to the Marxist cultural–historical narrative schemas that Bourdieu challenged. Far from it. The problem rather concerns the respects in which the Kantian armature that underlies these continues to inform Bourdieu's own narrative schemas and the impact these have on the form of 'guided freedom' he aspires to offer. In the differential architectures of the person that he proposes; in the unequal distribution of the capacity for freedom that this gives rise to; in the tutelary role that he assigns to the collective intellectual: in all of these ways, Bourdieu's work bears testimony to those forms of liberal political reasoning that have informed the relations between making culture and changing society within the institutions of the culture complex.

Afterword

In his critical engagement with Latour's work, Graham Harman draws attention to the moment of revelation that Latour records in *The Pasteurisation of France* when, as a young man driving through Burgundy, he first articulated what would come to be the founding principle of his philosophy. 'Nothing', Latour writes, recalling this moment, 'can be reduced to anything else, nothing can deduced from anything else, everything may be allied to anything else' (Latour 1988: 163). The passage provides a useful summary of the perspectives – pithily articulated by Latour, but which have a wider currency – in contemporary social theory that I have drawn on in this book. The first two parts of the sentence condense the reasoning that has inclined recent thought in favour of 'flat ontologies of the social' and against those dualisms (culture/society) that – whether in a reductive or a deductive movement – seek to account for one side of such divides in the terms provided by the other. It is, however, the third part of Latour's 'Pauline' roadside revelation that provides the perspective I have drawn on in the account of culture that I have offered. Its implication, as Harman interprets it, is not merely that of calling into question the division between human and non-human actors. It also de-substantialises the world, denying the existence of any pre-given essences, or essential divisions between this and that, in favour of an account of how realities are produced and (always provisionally) stabilised, through the alliances that actors enter (and are entered) into with one another. Particular forms of power; the arrangement of actors into particular assemblages; the ordering of the relationships between such assemblages: these are all the results of the processes – some of them long historical ones, others quite short; some with an extended territorial reach, others quite local – through which things, persons, techniques and technologies are brought into alliance with one another in particular configurations whose status is secured as that of a public ordering of things or, as Harman puts it, a 'public performance in the world' (Harman 2009: 66).

It is in this light that I have proposed that we should approach the current partitioning of the relations between culture and society as the publicly instituted and enacted outcomes of such processes. I have, with regard to culture, paid particular attention to the role of a historically particular set of knowledge practices – conceived, not abstractly, as the activity of Mind, but as always materially entangled in assemblages of various kinds – in producing new kinds of cultural actors in the world, and in organising their social deployment through the 'working surfaces on the social' they generate. My particular concerns in these regards have focused on how such cultural knowledges and assemblages have been at work in organising and distributing particular kinds of freedom. This has involved a consideration of the modes of production and deployment of both anthropology and aesthetics across a range of metropolitan and colonial settings; of variations within these according to whether their relations to conduct are routed via the public or the milieu; and an examination of how such knowledges and the forms of authority they produce have constituted particular forms of intervention into the regulation of conduct through the architectures of the person they organise.

My examples of the operations of the culture complex in these regards have been largely limited to state institutions. They have also been mainly drawn from a period stretching roughly from the mid-nineteenth to the mid-twentieth century. While I have provided examples indicating how the terms of the analysis I have proposed are more generally applicable – to the reality shows of commercial television networks and to the contemporary diagram of tolerance, for example – it is clear that these would need to be adapted and, no doubt, revised in any adequate application to the significantly changed intellectual, material and infrastructural conditions of the present. My attention has also focused predominantly on the ordering practices of cultural knowledges and institutions, and on the governmental rationalities informing the modes of their social deployment. The perspective of ontological politics makes it clear, however, that such ordering practices are always partial and incomplete; that there are always clashing knowledges – official and unofficial – in play, with discordant and messy effects; and that the real is made up of such criss-crossing flows. A fuller pursuit of these matters will, however, have to await another occasion.

Notes

1 After culture?

1 See, for a collection of essays bringing the perspectives of assemblage theory to bear on the concerns of cultural analysis, Bennett and Healy (2011).

2 There has yet to be a systematic appraisal of the relations between Bourdieu and Anglophone cultural studies, although see Ahearne (2010) for useful pointers to key differences. There were significant interrelations between them in the 1960s and early 1970s: Bourdieu included Richard Hoggart's (1969) *The Uses of Literacy* in the book series he edited and personally oversaw its translation (Long 2008); he thought highly of Paul Willis's (1977) *Learning to Labour*; and he invited Raymond Williams to lecture at the École normale while also publishing his work in *Actes de la recherche sociologique*. Williams, in turn, did much to introduce Bourdieu's work to English-speaking readers, alongside Richard Nice, Bourdieu's main English translator and a member of the Birmingham Centre for Contemporary Cultural Studies during the period of Stuart Hall's directorship (Bourdieu and Wacquant 1992). Thereafter, however, Bourdieu adopted a more distanced relationship to cultural studies that, in its concern to validate the resistances of subordinate subjects, increasingly resembled the positions – of De Certeau, for example – that Bourdieu resolutely opposed in France.

3 I address the similarities and differences between the foundational texts of British cultural studies and Bourdieu's work in these regards in Bennett (2011b).

4 I include here Howard Caygill's (1989) systematic dissection of Kant's aesthetic and, more generally, Ian Hunter's (2001) critical assessment of the relationships between Kant's metaphysics and the rival tradition of civil philosophy. I discuss these more fully in Chapters 6 and 7 respectively.

5 Where, however, his work has also been subjected to critical assessment. Meaghan Morris, one of the first key figures to introduce Deleuze and Guattari to Anglophone cultural studies, also assesses the limitations of their nomadology from a feminist perspective: see her essay 'Crazy talk is not enough' collected in Morris (2006: 187–201).

6 This has been especially true of Bruno Latour's contribution to what is now a quite widespread interest in reviving the rival sociology of Gabriel Tarde: see, for example, the essays collected in Candea (2010) and the special issue of the journal *Economy and Society* (36, 4, 2007) edited by Andrew Barry and Nigel Thrift.

7 See, for contrasting discussions of the relations between Foucault and Bourdieu, Callewaert (2006) and my response to this in Bennett (2010a).

8 I am indebted to John Frow, in his comments on earlier collaborative work, for this observation.

9 This is an over-simplification that does not take adequate account of Bourdieu's introduction of field theory into sociology.

10 Latour (2010) has since developed this perspective into a manifesto. There are similarities between this focus on the compositional makeup of social life and the 'topological turn' in the social sciences. I prefer the former in view of the stress it plays on such compositional qualities as the outcomes of agential processes.

11 Rose has objected that the relations we have to ourselves cannot be understood by locating them in 'some amorphous domain of culture' but have to be approached from the perspective of government. There is, however, as I have argued previously (Bennett 1998), no need to conceive culture as an amorphous domain but rather, and in precisely the same way Rose proposes for the psy sciences, as an assemblage of governmental techniques.

12 That said, Rose's essay on the death of the social (Rose 1996) still offers the best means of engaging theoretically with the shift to 'community' as a cultural resource for governing.

13 References to this source are not to pages but to locations in the Kindle edition.

2 Making culture, organising freedom, changing society

1 My concern, in interpreting aesthetics as a cultural knowledge, is not with the cognitive, sensory or affective effects of artistic practices but with the accounts of such practices proposed in the tradition of Western aesthetic discourse running from the third Earl of Shaftesbury through to Jacques Rancière.

2 This argument has since been applied more generally to a wider range of cultural institutions by Chris Otter's (2008) account of oligoptic vision.

3 See, for a more extended critical engagement with the concept, Hall (2006). I have also reviewed the limitations of the concept in Bennett (2006). I draw on these discussions here.

4 Conal McCarthy (2007) demonstrates a very early history of engagement with museums as important spaces for Maori self-representation.

5 I have addressed these shortcomings in Bennett (2004, 2010 and 2012a).

6 I stress, though, that I do not, as does Ray, endorse this essentially Kantian conception of culture on its own terms. The aspect of Ray's discussion that I most value is the tension it sets up between, on the one hand, culture as a historically specific mechanism for the shaping of identities and, on the other, the everyday customs, values and traditions that comprise the ground on which culture works. That said, Ray's account of the ways in which this mechanism works are a little too abstractly and generally stated. The result is an over-unified account of culture as always working through the same operations rather than, more convincingly, an *ad hoc* assemblage of different machineries for the shaping of identities and conducts.

7 Bruce Curtis (2002) points usefully to a series of ambiguities and contradictions in Foucault's discussion of these matters. Defining 'population' sometimes as populousness, sometimes as the social body, and sometimes as a statistical aggregate, Foucault locates its emergence as an object of governance sometimes in the eighteenth century and sometimes a century later. Danica Dupont and Frank Pearce (2001) identify a number of troubling inconsistencies between the principles of historical reasoning underlying Foucault's earlier archaeological method and the tendency towards a form of Hegelian objective idealism that they detect in his account of the historical unfolding of the relations between government and population.

8 See also Foucault's earlier discussion of milieu, in *Society Must Be Defended* (2003), couched largely in terms of the difference between the individualising orientation of discipline and the massifying qualities of biopower. Foucault is very clear in this discussion that, by milieu, he means man-made as well as natural environments and, crucially, the relations between the two.

9 For a more detailed discussion of the relationship between speciation, security and biopower, see Dillon (2004).

10 The contrast Foucault evokes here draws implicitly on the dazzling spectacle of royal palaces that provides Immanuel Kant (1987: 45–6) with the image of the stupefying effects of sovereign power to which he juxtaposes the critical interiority of the aesthetic.

11 I have argued this case more fully in Bennett (1999). See also, for a related account of the governmental production of the public sphere in a colonial context, Kalpagam (2002).

12 Kalpagam (2002a) is similarly instructive on this point.

13 I touch here on a complex set of questions concerning the relations between the rise of the social, the deconfessionalisation of politics and secularism. Keith Baker has suggested that the conception of the social as an autonomous and determining ground of existence 'was unthinkable within a religious imaginary that saw the entirety of order and existence – metaphysical and physical, natural and human – as emanating from the Divine' (Baker 1994: 195). Baker sees the deconfessionalisation of politics as crucial for the later emergence of a governmental relation to population that, he argues, required the existence of a domain of the social that is 'assumed to possess its own regularities, autonomous life, and independent existence' (197). Once, in varying ways and to varying degrees, religion and state have been separated in order to disconnect political and administrative affairs from the disordering effects of religious factionalism, it was then possible to invoke religion as a supplementary moral force capable of acting on the social so as to secure order. The service that religion is then called on to perform is one that subordinates it to the requirements of a society that is presumed to exist independently on its own foundations. 'Religion', as Baker puts it, 'is ultimately justified in the name of society, and not vice versa' (Baker 1994: 107). This schematisation of the state–society relation also informs the organisation of the governmental space in which secular forms of cultural knowledge have operated, sometimes in tandem with and sometimes in opposition to religions.

14 I stress these points here as a counter to Mitchell Dean's (2007) characterisation of a concern with cultural governance as concerned exclusively with abstracted moral or ideational forces.

15 See also, for example, Carroll (2006), Otter (2007) and Crook (2007).

16 See also on this matter Mary Poovey's (1995) classic account of the role of visual metaphors in the production and division of the social body in nineteenth-century England.

17 See Legg (2007) for a discussion of similar processes at the city level.

18 I draw here on my more detailed discussion of these questions in Bennett (2009).

19 For fuller elaborations of this argument, see Bennett (2007, 2007a).

20 I offer this example as similar to one Latour gives at a later point in his discussion (Latour 2005: 40).

21 Ogburn draws here on Latour and Woolgar's (1986) analysis of laboratory practices.

22 See on this Philip Connell's (2001) reassessment of the relations between Romanticism and political economy.

23 Although Boas played a significant role in challenging racial conceptions on the basis of cultural conceptions of difference, his work contained residues of the former. His work was also problematic in bringing a different perspective to bear on the politics of difference insofar as these concerned Native Americans and African Americans, applying a preservationist logic in relation to the former and an assimilationist one to the latter. See Baker (2010).

24 For fuller discussions, see Bennett (2007, 2007a).

25 Noel Pearson has articulated this position most consistently in his regular column for *The Australian*. A particularly clear example is his commentary on the report of

the Expert Panel on Constitutional Recognition of Indigenous Australians (Pearson 2011). While supporting the removal of the race clause that provided the basis for the inclusion of Aboriginal people in the Constitution after 1967, Pearson urges the need for the Constitution to recognise a special place for Aboriginal people on the basis of their cultural and historical status as indigenous. This is necessary, he argues, to provide the basis for special forms of governmental action to be brought to bear on the conditions of Aboriginal Australians – including those that work through the carrot-and-stick mechanisms of the NTI – if the gap in wellbeing between Aboriginal and white Australia is ever to be closed.

26 For an illuminating account of the changing role of archival practices in these processes, see Spieker (2008).

27 Bourdieu identifies the differences between his and Luhmann's approach in Bourdieu and Wacquant (1992: 103).

28 Bourdieu reaches a similar conclusion in his ill-considered assessment of Latour and Woolgar's account of laboratory work. Their account of the roles of writing, translation and inscription in scientific research, he argues, constitutes a '*semiological vision of the world* which ... leads them to that paradigmatic form of the scholastic bias, *textism*, which constitutes social reality as a text' (Bourdieu 2004: 28).

29 Law and Urry cite the work of Osborne and Rose (1999) on public opinion. However, this line of argument was developed much earlier by Pierre Bourdieu in his probing study of the ways in which public opinion research produced its own reality effects: see Bourdieu (1979).

3 Civic laboratories

1 I place 'natural objects' in quotes since, as Frow (2004: 358–61) shows, there can be no clear-cut distinction between culturally defined and naturally occurring objects. The 'naturalness' that is at issue here then concerns those objects that laboratory practice takes as natural.

2 There are clear connections between Knorr-Cetina's arguments here and Foucault's earlier analysis of the relationships between medical practices and the clinic (Foucault 1973).

3 Christopher Whitehead's (2009) account of the relations between art and archaeology in British museum practice from the 1850s to the 1880s offers a detailed account of the different ways in which these two knowledge ensembles assembled and exhibited material culture while simultaneously operating as parts of social processes of ordering, which distribute differential statuses between both the experts who organise them and the publics who visit them. I have discussed this in more detail in Bennett (2012a).

4 This aspect of Latour's conception resonates with Derrida's contentions regarding the political significance of archives. The intellectual ordering of the materials archives assemble, Derrida (1998) argues, is crucially implicated in the exercise of social order and authority.

5 Belting develops his conception of the invisible masterpiece via a commentary on Balzac's 1845 short story 'The unknown masterpiece': see Balzac (2001)

6 The nineteenth-century literature contains many examples of arts administrators and museum directors who were prepared to sacrifice aesthetic complexity for the accessibility of more familiar kinds of art as the price of enlisting art in the cause of improving public manners. This was especially true of Henry Cole whose reasoning on this matter I have discussed elsewhere (Bennett 1992).

7 The individualising aspect of this strategy is important and in sharp contrast to earlier forms of Aboriginal administration that aimed to 'civilise' Aborigines as members of segregated communities. Aboriginal participation in such experiments was based on the mistaken belief that the application of Lockean principles would

eventually earn them the right to community-based forms of freedom and self-governance in recognition of the rights derived from the application of their labour to the lands granted them. See Attwood (2004).

8 See, for example, the contributions – including my own – in Lumley (1988), an early and influential collection of the 'new museology' in English.

9 Thompson, though, goes on to take actor–network theory to task for subscribing to too unbounded and open-ended a conception of the capacity for networks to be endlessly rearticulated.

10 This is true, in some degree, of Eilean Hooper-Greenhill's *Museums and the Shaping of Knowledge* (1992) and of my own *The Birth of the Museum* (Bennett 1995).

11 Kevin Hetherington's discussion of the work of museums as places where the 'waste matter' of commodity culture is disposed of as a practice of incessant reordering points in a similar direction (Hetherington 2007: Chapter 7).

12 I have discussed the range of positions on the relations between museums and cultural diversity more fully elsewhere: see Bennett (2006).

13 This was an especially evident aspect of Neil MacGregor's address in a public seminar held to mark the opening of the *Enlightenment* exhibition. Organised around the contemporary role of universal survey museums, the seminar was addressed by the directors of five such museums – the Louvre, the Hermitage Museum, the National Museum in Berlin, the Metropolitan Museum of Art and the British Museum. Macgregor invoked the *Living and Dying* exhibition as an example of the museum's commitment to diversity, and a way of exhibiting diversity that would not be possible but for the ways of accumulating and storing the world developed by Enlightenment forms of collecting.

4 Making and mobilising worlds

1 Spencer, a product of Mancunian liberalism, was originally trained as a natural historian and was an early supporter of Darwinism. His substantive position at Melbourne was as professor of biology but his interests progressively shifted to anthropology. Spencer met Gillen, a postmaster and Justice of the Peace at Alice Springs, during the Horn expedition. They subsequently developed a lifelong collaboration in which the distinction between the roles of field collector and university-based interpreter, while remaining intact, was also significantly blurred.

2 I shall follow Wolfe here in retaining the orthography of Spencer and Gillen, rather than using the corrected form of the Arrernte, since my concern is with the operation of this category in nineteenth-century Eurocentric discourses.

3 It is these issues, I think, that Howard Morphy (1996 and 1996a) loses sight of in castigating Wolfe as a post-modernist constructivist whose interpretation of the dream-times as entirely a property of white discourse fails to appreciate how they were 'an integral part of the conceptual structure of Aboriginal society, the recognition of which was part of a process of value creation that ran counter to the colonial process' (Morphy 1996: 164). In Wolfe's analysis, the later take-up of the language of the dream-times by different Aboriginal communities constituted a form of acquiescence to the terms in which Aboriginality had been encoded in anthropological discourse. Morphy's response to this is to suggest that Aboriginal take-up of the term was prompted by the fact that it 'fitted a lexical gap in Aboriginal languages' (Morphy 1996: 178). It allowed a semantic field that stressed the primacy of ancestral beliefs and values to the organisation of Aboriginal modes of sociality that was prominent in most Aboriginal languages to achieve a common currency across tribal groupings in ways that were of salience to Aboriginal people in the context of anti- and post-colonial political mobilisations. Although Morphy thus clearly wants to stake out a different political position from Wolfe by stressing

the agency of Aborigines, this is still an agency exercised in the context of a new set of realities and relationships that had been historically constructed in the relations between coloniser and colonised. The key issues concerning the dream-times and their successors in this regard are not whether they were or were not realities that pre-exist(ed) their inscription in anthropological discourse but, as John Law (2004: 122–42) suggests, the functioning of the different ways in which they have operated within the different method assemblages – ways of producing and enacting truths – within which the relations between black and white Australia have been complexly enacted.

4 Although they do not use the terminology I have proposed here, a similar understanding of Aboriginal culture informs a number of recent studies including Pat O'Malley's (1998) account of how indigenous practices of resistance have shaped practices of colonial rule as well as being shaped by them; Bain Attwood's (2004) account of the mobile constructions of 'Aboriginality' that have informed both state practices of indigenous governance and the development of Aboriginal rights and political movements; Chris Healy's (2008) account of the ways in which the strategic forgetting, and subsequent re-remembering, of Aboriginal culture has mediated the relations between 'mainstream' and indigenous Australians; Tim Rowse's (2009) account of how the statistical assembling of indigenous populations in official censuses constitute new transactional realities that are mobilised as parts of the processes of collective subjective formation on the part of Aboriginal political movements; and Fred Myers' (2002) account of the relations between Aboriginal and Western art practices and institutions.

5 The separation of the fieldwork phase from earlier moments in the history of anthropology is contentious. This phase was itself far from unified and, as numerous historians of anthropology have argued, bore significant resemblances to earlier forms of voyaging into the Other and the literatures these produced (Debaene 2010, Defert 1982, Fabian 2000). The so-called immersive paradigm that allegedly distinguished fieldwork was also, Anthony Pagden (1986) has argued, amply rehearsed by missionaries in early Latin American colonial contexts. My remarks here, then, need to be read in terms of the more limited set of differences concerning the relations between sites of collection and centres of calculation from those associated with 'armchair anthropology' rather than as validation of the claims made on behalf of fieldwork as marking the transition between anthropology's scientific prehistory and its establishment on a properly scientific footing. Henrika Kuklick (2011) offers a very considered discussion of these distinguishing features.

6 Alfred Haddon's role in relation to the Cambridge 1898 Torres Strait expedition and the 1902 period of fieldwork among the Todas by W. H. R. Rivers (who had also been a member of the Torres Strait expedition) are the examples Morphy cites, although Franz Boas' participation, while at the American Museum of Natural History, in the 1897 Jesup North Pacific Expedition is also relevant. See Herle and Rouse (1998) for discussions of the Torres Strait expedition, and Rexer and Klein (1995) for the Jesup expedition.

7 The degree to which this was so, however, varied. In contrast to the case represented by Spencer in bringing together the roles of fieldworker and museum curator/administrator, the Musée de l'Homme, discussed in the next chapter, represents a different and more dispersed set of relations in which these functions were spread across a range of actors. These are not inconsequential considerations. We have now a range of studies that convincingly demonstrate the significance of the routes and mechanisms through which objects travel to museums (see Alberti 2009, Gosden and Larson 2007, Jones 2007)

8 The role of Spencer's work on the Arunta in shaping Durkheim's 1915 account of the elementary forms of religious life (Durkheim 1964) has been subject to extensive commentary and is discussed by both Wolfe and Morphy.

9 This is truer of British anthropology than it is of North American anthropology in general and of Franz Boas in particular. For Boas, the use of sound recording devices constituted an indispensible means for accumulating a material archive of Native-American languages that would provide an evidentiary basis for the anthropological study of such languages that would match the sources available to classical philology (see Briggs and Bauman 1999, Jacknis 1996).

10 The consequences flowing from the inclusion of Aboriginal 'art' in art exhibitions has been extensively studied in this regard. See Griffiths (1996) for a discussion of the ripple effects of the Melbourne 1929 Aboriginal Art Exhibition and Myers (2002) for the whole series of adjustments and counter-adjustments of government policy agencies, international art markets, Aboriginal artists and cultural intermediaries, etc. to one another flowing from – aptly enough – the *Dreamings: The Art of Aboriginal Australia* exhibition of acrylic dot painting at the 1988 Asia Society Galleries in New York.

11 The distribution of rations had often been used as a means of managing relationships across the colonial frontier on a private basis. What was new after 1880 was the supplementation of the philanthropy of police and pastoralists in this regard by government-run ration depots (Jones 2007: 369).

12 I have discussed these elsewhere: see Bennett (2004).

13 The meeting was attended by Bronislaw Malinowski on his first visit to Australia (Stocking 1991: 35).

14 He does not, though, regard the Aborigine as a pure race, seeing it as a probable blend of earlier black races.

15 The qualification is important. Spencer acknowledges that the introduction of Reckitt's Blue into Australia by the white man led to an acquired capacity to recognise blue, leading to its use in Aboriginal art. This is then re-presented as the proposition that the use of blue in any Aboriginal artefact is sufficient proof that it is a corrupted form: 'the presence of a blue pigment on any Australian ornament or implement that finds its way into one or other of our museums is regarded by all Curators as clear proof that the tribe from which it comes has lost its primitive outlook on art' (Spencer 1921: 4).

16 The passage is from a paper that E. B. Tylor presented on Howitt's behalf to the Anthropological Institute in Britain in 1883.

17 See, for example, Gary Wilder's account of the part played by the practices of 'colonial humanism' associated with the inter-war phase in the development of French anthropological fieldwork in ordering the relations between France and its African colonies (Wilder 2005).

18 These were the two judgements, made in 1992 and 1993 respectively, which overthrew the doctrine of *terra nullius*, the legal fiction that Australia was unoccupied at the point of white settlement.

19 There is, though, an exception in Spencer and Gillen's *Across Australia*, which includes one picture of two clothed Aboriginal women. The accompanying text states that the Aborigines of Central Australia went 'stark naked' and had no idea of clothing until coming into contact with the white man, accepting government blankets and cast-off clothing from settlers. They did not, however, acquire the habit of wearing clothing, passing it on from one person to another, so that, sometimes being clothed and at others going naked, they became liable to attacks of phthisis as one manifestation of racial deterioration brought about by contact with whites. See Spencer and Gillen (1912: 186).

20 I use the terms of this racial vocabulary ('half-castes', 'full-blood', etc.) rather than the contemporary vocabulary of 'mixed descent' as the language that belonged to – indeed, constituted – the historical realities I am concerned with here.

21 It is important to note that Spencer was the first to serve in this capacity after, in 1911, responsibility for the administration of Aborigines in the Territory was transferred to the Commonwealth Government.

22 See McGregor (1997) for a detailed discussion of these.

5 Collecting, instructing, governing

1 The history of the relations between the terms 'ethnography' and 'ethnology' in France is a complex one. While often overlapping in their meaning and uses, ethnology was originally associated with physical anthropology and was strongly secular and materialist in orientation while ethnography also had affiliations to spiritualism. It acquired strong connections with folklore studies in the opening decades of the twentieth century. In the 1920s and 1930s, particularly after the establishment of the Institut d'ethnologie in 1925, the term 'ethnology' came to encompass both a more scientific conception of ethnography, largely in the Maussian tradition, and a continuation of the earlier physical anthropological conceptions of the term. Rivet played a crucial role in brokering and maintaining this synthesis in his positions at both the Institut and the MH.

2 The ATP did not, however, open in its own premises until 1972; see Segalen (2005).

3 This is especially true of the MH, which, so far as I have been able to determine, still lacks a comprehensive account of its history from its early formation through to the present of the kind that Segalen (2005) provides for the ATP.

4 New Guinea was the chief exception here. Previously a German colony, New Guinea was established as a mandate territory under Australian administration under the terms of the partitioning of Germany's colonies that was undertaken by the League of Nations after the 1914–18 war. See Sibeud (2007) for details.

5 This is an oversimplification of a complex history in which colonial populations were themselves differentiated in terms of their access to civic rights – differentiated both across different colonial contexts and, in some cases, within the same colony – on a mixture of racial and status grounds. I refer the reader to Conklin (1997) and Wilder (2005) for more detailed discussions of these questions.

6 This is the general, and well-documented, line of argument developed by the editors and contributors to Blanchard *et al.* (2008).

7 The relations between these two orientations are discussed by Conklin (1997) as they are also, albeit from a somewhat different theoretical perspective, by Dias (2010).

8 Initially, in 1931, the Musée permanent des colonies and, from 1935, the Musée de la France d'outre-mer.

9 Initially the Laboratoire d'Ethnographie Française, directed by Maget, this subsequently became the ATP's Centre d'Ethnologie Française.

10 I draw, in the following discussion, mainly on Blanckaert (1988), Fabre (1997), Sherman (2004) and Sibeud (2007).

11 Grognet (2010) attributes considerable importance to Rivet's location in the Muséum national d'Histoire naturelle in ensuring that the work of the MH complied with significant aspets of the Muséum's long-term intellectual and organisational agendas. Rivière, Maget and all the other MH curatorial staff worked *in situ* at the Palais de Chaillot.

12 Not to be confused with the later cultural historian perhaps best known to English-speaking readers as the author of *Portrait of the King* (Marin 1988).

13 Lebovics (1992) is particularly helpful in dissecting the terms in which these rival left and right-wing versions of anthropology vied with one another in their endeavours to influence the frameworks in which the identities of the different populations of Greater France would be defined and governed. Marin had proved influential in the Chamber in delaying the financial support needed to put Rivet's reform programme at the MET into effect.

14 For fuller discussions of the earlier French tradition of physical anthropology and its relations to museums practices, see Dias (2004) and Dias (1991), respectively.

15 This conception of the MH's function distinguished Rivet's conception of ethnology from Mauss' conception of ethnography as the basis for a science of man in society that was to provide a universal account of the development of human societies

from mechanical to organic forms of solidarity. For Rivet, by contrast, ethnology proposed a different kind of synthesis in aiming to unite somatic anthropology with ethnography and linguistics to establish a science of the human species, its branches, origins and differentiations. It was in this respect, Grognet argues, that Rivet's controlling vision for the MH derived from, and asserted the continuing influence of, the comparative anatomy orientation of the Muséum national d'Histoire naturelle (see Grognet 2010: 280–84, 430–34). At the same time, however, as Jamin (1998) stresses, Rivet's project was for a synthesis of the natural and social sciences of man, in the Enlightenment tradition of the Société des Observateurs, rather than for the subordination of the latter to the former.

16 See 'Un rapport signe par Paul Rivet sur une demande de subvention a l'outillage national qui presente un projet de restructuration de Musée', 21 April 1934, Archives of the Musée de l'Homme, 2 AM 1 G3b: *Musée d'Ethnographie: notes et raports, activité* (1934).

17 See 'Notes relating to the activity of the Museum of Ethnography in the French colonial domain', Archives of the MH, 2 AM 1 G3b: *Musée d'Ethnographie: notes et raports, activité* (1934).

18 It is, indeed, doubtful whether any of the missions organised by the MH could be described as typical. Grognet's discussion of the most important missions in the annexes to his thesis reveals a remarkable diversity in the ethos, orientations, purposes and practical arrangements of these missions, sufficiently so to suggest that expediency often trumped the MH's aspiration to develop a clear overriding scientific rationale for its fieldwork activities.

19 Malinowski invoked the image of the anthropologist stepping outside the 'closed study of the theorist' and down from 'the verandah of the missionary compound' into the 'open air of the anthropological field' (cited in Stocking 1983: 112).

20 Bondaz (2011), in a recent useful reminder of the MH's connections through Rivet to the Muséum national d'Histoire naturelle, draws our attention to fact that the Dakar–Djibouti expedition was also mandated to collect natural specimens, both living and dead, and to the respects in which many of its activities – including those geared towards the collecting of ethnological materials – were imbued with the culture of the hunt.

21 The American anthropologist credited as one of the forerunners of anthropological fieldwork for his work among the Zuni in the 1890s.

22 For an early and formative summary of these principles, see the document prepared by Anatole Lewitsky 'Quelques consideration sur l'exposition des objects ethnographiques', Archives of the Musée de l'Homme, 2AM 1 G3d, *Notes and Reports, 1935–7.*

23 For examples of this correspondence, see the MH Archives 2 AM 1 M2c *Mission Dakar-Djibouti, 1931–33 Correspondence.*

24 Éric Jolly (2001) offers a good account in this respect of the relations between the Dakar–Djibouti mission and the 1931 International Colonial Exhibition. The departure of the mission followed closely on the opening of the Exhibition, and its activities were widely reported throughout its duration as contributing to the Exhibition's celebration of colonial possessions as a part of its promotion of the identifications of Greater France.

25 See *Paris-Midi*, 20 November 1930. Copy held in the MH Archives, file 2 AM 1 B9, *Mission Dakar-Djibouti.*

26 I am indebted here and in other aspects of my discussion of this mission to Clifford (1983).

27 The MH was partly funded by the ministries responsible for France's colonial and overseas territories, and its opening was fan-fared by colonial troops.

28 Rivière's formulations, which pretty much echoed those of Mauss, were, in their turn, echoed by M. Roustan, the Minister of Public Instruction (Archives of the

MH, 2 AM 1 B9, *Mission Dakar-Djibouti*) and by Griaule when interviewed by *Le Jour* in connection with his later (1935) Sudan expedition (Archives of the MH, 2 AM 1 B8 MET/MDH, *Missions Ethnographique*).

29 That said, there have been objections on these grounds too. Jamin (1986), for example, argues that *Minotaure* only became a surrealist publication after the issue in which the Dakar–Djibouti mission was reported.

30 The surrealists mounted significant critiques of the Colonial Exhibition of 1931, timed to celebrate the centenary of French rule in Algeria. They ran public campaigns against visiting the exhibition and took issue not only with the history of French colonial massacres and exploitation but also with the concept of Greater France, which they judged an 'intellectual swindle' designed to delude the citizens of the metropole into believing themselves imperial proprietors by adding to French landscapes 'a vision of minarets and pagodas' (cited in Lebovics 1992: 56). They also organised their own Anti-Imperial Exhibition, which critiqued the status accorded ethnographic collections in Western collecting institutions.

31 See on this Pierre (2001/2).

32 It would take me too far away from my concerns to pursue the point in detail, but it should be noted that Clifford's conception of 'ethnographic surrealism' depends a good deal on his interpretation of *Documents* as a key site of convergence between ethnography and surrealism in the 1920s, particularly with regard to the 'shock value' he attributes to both. Denis Hollier and Leisl Ollman (1991) suggest this is misplaced to the degree that ethnography and surrealism attributed quite sharply different meanings to the concept of the document derived from different disciplinary histories of use and interpretation.

33 I argued at the beginning of this chapter that the distinctiveness of both the MH and the ATP belongs largely to their formative years. This was largely because, soon after their official establishment, each was faced with the difficulty of finding some accommodation with, respectively, the racial and regional politics of the Nazi occupation. These issues were worked through differently by the two institutions. It would be a mistake, however, to read off the different accommodations they offered from the fates of their two directors: Rivet, going into exile in 1941 and retaining an untarnished reputation as a leading figure of the Resistance in which many MH staff were active while Rivière, remaining in post throughout the war, was subsequently accused of collaboration with the Vichy government. There is a large literature on this question. The most persuasive analyses are those that de-personalise the issue to focus on field dynamics – that is, on the respects in which the relations between the French and German scientific fields in the domains of ethnology and folk studies reactivated divisions within the former with significant consequences for the positions that were open to Rivière – rather than on questions of individual political ethics. See, for example, Fabre (1997), Lebovics (1992) and Weber (2000).

34 This dismantling – the MH was closed in 2003 for (at the time of writing) a still-ongoing process of reorganisation – was followed two years later by the closure of the ATP, which, from 1972, had its premises in the Bois de Bouloigne.

35 Again, there is a large literature on these questions: see, for example, Clifford (2007), Desvallées (2007) and Price (2007).

6 The uses of uselessness

1 Osborne's choice of Yúdice as the main target of his criticism (he also singles out my work for attention) is somewhat surprising since Yúdice is critical of many of the tendencies towards 'the expediency of culture' that he describes.

2 I should add that this approach is also at odds with my assessment – quite some time ago now – of aesthetics as a useless form of knowledge (see Bennett 1985; also in Bennett 1990). While still, I think, valid in some respects, this lacks an adequately

nuanced appreciation of the historically variable relations between uselessness and freedom and the enormous productivity of these with regard to the actual uses of art that they have engendered.

3 By defining every individual in terms of his relation to production, Petty's political arithmetic, developed initially in connection with the tasks of colonial government in Ireland, meant that every individual could be measured according to the same universal equivalent by which the productivity of land was measured. This produced a ground on which the Irish and English could become like one another, as instances of *homo economicus*, with religious preference, culture, language and political affiliation becoming no more than 'Perversions of Humour'.

4 There are, though, close relations between these two aspects of the disinterestedness of aesthetic judgement. I have discussed how these interact in Bourdieu's work in Bennett (2007c).

5 Catoptric vision refers to a device invented by the Mersenne circle that obviated the need for the spectator to take up a position to one side of the painting in order to decipher anamorphic works of art. It served Hobbes as a means of visualising the production of political order from the single and unchanging position of sovereignty (Caygill 1989: 19–25).

6 I draw here on Guyer's discussion of Kant's position in his 1775–6 course of lectures on aesthetics (Guyer 2005: 175).

7 There is a large literature on this topic. In addition to the work by Dowling (1996) already cited, see Goodlad (2003), Thomas (2004) and Woodmansee (1994). See also the very useful historical survey of the implications of different conceptions of the aesthetic for the uses to which art might be put offered by Belfiore and Bennett (2010).

7 Guided freedom

1 For a fuller discussion, see Bennett, Savage, Silva, Warde, Gayo-Cal and Wright (2009).

2 See O'Connor (2010) for a good account of the respects in which this is so in relation to the development of rock music in the 1960s and 1970s.

3 See Bennett, Savage, Silva, Warde, Gayo-Cal and Wright (2009: 86–8).

4 Here and throughout I draw on the helpful discussions of Rancière's concept of police offered by Samuel Chambers (2011) and Jodi Dean (2011).

5 Rancière (2011) helpfully distinguishes his position in these regards from resistance theory and from Giorgio Agamben's conception of politics.

6 The concept of metapolitics, introduced briefly in *Disagreement*, is accorded a more central significance in later texts: see Rancière (2002, 2004a, 2009). I have discussed these in more detail in Bennett (2010c).

7 See Toscano (2011) for a good discussion of Rancière's position on sociology and a critical assessment of its limits.

8 Rancière makes no mention of what is arguably the most important aspect of Kant's discussion here. For it is not merely disinterest in the *uses* of an object that Kant makes the defining attribute of judgements of the beautiful; it is also, and more importantly, disinterest in the object's *existence*. This transforms taste into a question of the activity of subjects in relation to the presentation of objects within their selves quite independently of the object's existence. The significance of this move in Kant's conception of the aesthetic as a domain of self-regulation conducted within the interiority of the subject goes unremarked by Rancière. Yet it is crucial to an understanding of the relations between aesthetics and modern practices of liberal government.

9 This disputation of the Marxist account of the radicalisation of the working class as arising from its proletarianisation as an outcome of the process of industrialisation is the primary political point of Rancière's account of working-class nocturnal reading habits: see Rancière (1989).

10 Rancière places particular emphasis on Auguste Blanqui's 1832 declaration of the proletariat as the profession of those who live by their labour but who are deprived of political rights – a scene he returns to again and again.

11 Jonathan Rose's (2001) telling historical recovery of the role played by Kantian conceptions of the relations between aesthetics and freedom in mid-twentieth-century adult education movements merely confirms how far contemporary working-class practices are removed from such conceptions.

12 The endeavour of the *Annales* school to transform history into a social science discipline is the subject of Rancière's critique in Rancière (1992).

13 I refer here to the account of the schoolmaster Jacotot in Rancière (1991).

14 In much of what follows I take a leaf out of Tom Bowland's (2007) argument that critique, in its variant forms, does not merit being treated as an exception to the forms of intellectual authority it takes issue with.

15 See on this Reinhardt Koselleck's discussion of the history of the concept of *Bildung* in Koselleck (2002).

16 See on this Dias (2004) and Nakata (2007).

17 My reference here is to the title of the Tate symposium at which I presented the paper from which this chapter derives.

18 I hesitate over the use of the concept of globalisation considered in its relations to art and collecting practices for reasons that I have discussed elsewhere: see Bennett (2006).

19 I draw here on Howard Morphy's discussion of Yolngu art in Morphy (2007).

20 See Hunter (1991) and du Gay (2000) for further elaborations of this point.

21 I have discussed these questions in Bennett (2006a).

8 Habit, instinct, survivals

1 I draw here on James Tulley's (1989) probing discussion of Locke.

2 I am indebted here to the paper 'Raivaisson and phenomenology' presented by Mark Sinclair at the conference on *Habit as Second Nature* that was organised by the British Phenomenological Society at Oxford University in April 2011.

3 I note again that I follow convention here in using Spencer's spelling rather than the corrected form of the Arrernte, since my concern is with the operation of this category in nineteenth-century Eurocentric discourses.

4 I have commented on this remarkable text by Pitt Rivers on a number of previous occasions (Bennett 1995, 1998, 2004), and I draw on these discussions here, while also providing a new context for my reading of it by relating it to the discussion of the distinctive habit–instinct nexus sketched in above.

5 While some aspects of Pitt Rivers' account are similar to Darwin's account, published a year earlier (1874), of the relations between the intellectual faculties and habit in man and in the lower animals, the differences are more telling. Although Darwin also refers to Spencer and Lubbock, and although he admits that 'some intelligent actions, after being performed during several generations, become converted into instincts and are inherited' (Darwin 1952: 288), this is a long way from the general theories of the relations between habit and instinct characterising Spencer's neo-Lamarckian account.

9 Habitus/habit

1 There are connections between my argument here and my discussion of Bourdieu's concept of cultural capital in the light of Foucault's account of neoliberalism: see Bennett (2010a).

2 François Héran (1987) has argued that Bourdieu's contentions here place more weight on the distinction between habitus and habit than is warranted. The terms in

which Bourdieu distinguishes these depends a good deal on the difference between Latinate and vernacular forms, and thence on the difference between the values invested in sacred or scientific versus vulgar uses, which is not present in the medieval scholastic tradition since this was written entirely in Latin. Habitus was then simply habit. Or rather, habit then was more complexly habit than it subsequently came to be in the purely mechanistic interpretation that characterised its use in the Descartes–Kant lineage in the sense that it included a space within which more dynamic practices of self-formation might be developed.

3 I am interested in these questions too and have shown in detail why, on methodological grounds, Bourdieu's account of working-class culture in *Distinction* needs to be treated with a good deal of caution (Bennett 2011b).

4 I am indebted to Corcuff (2001) for drawing my attention to this aspect of Bourdieu's discussion.

5 Bourdieu presents a different position in his later work where he discusses the role of technical capital in the accumulation and differentiating strategies of sections of the working class (Bourdieu 2005).

6 Thoburn (2002) suggests a helpful qualification to this position by showing how, when Marx does allow the working class to act in its own right, this is only via the production of the *lumpenproletariat* as a position of absolute negation from which the working class can distinguish itself.

7 See the essay 'Reification and the consciousness of the proletariat' in Lukács (1971).

8 The criticism has a long history and one that is quite detachable from Rancière's politics. The earliest version I am aware of is that offered by Josef Révai (1971) in his review – first published in 1923 – of Lukács's work.

9 There are, though, equally occasions on which Bourdieu puts forward a contrary view as when, for instance, he argues that a sociological and historical understanding of the conditions underlying works of art – forms of understanding that, of course, depend on concepts – serves only to enrich our appreciation of those works.

10 I have discussed this attribute of post-Kantian aesthetics elsewhere; see Bennett (1990: 143–66).

11 I have discussed this aspect of Bourdieu's work in Bennett (2005).

12 He also notes the complexity of Bourdieu's position in relation to this tradition: always retaining elements of it in his insistence on the need for intellectuals to maintain a degree of autonomy in relation to the economic and political fields, Bourdieu was at the same time critical of those forms of ultra-leftism he analysed in *Homo Academicus* (Bourdieu 1988), lampooned as 'armchair Marxism' in an *Invitation to Reflexive Sociology* (Bourdieu and Wacquant 1992) and returned to in *Pascalian Meditations* (Bourdieu 2000) while, at the same time, reverting to a related oppositionalism in the various *contre-feu* of his late political career.

References

Adorno, Theodor (1963) *Critical Models: Interventions and Catchwords*, New York: Columbia University Press.
——(1984) *Aesthetic Theory*, London: Routledge & Kegan Paul.
——(1991) 'Culture and administration' in T. W. Adorno and J. M. Bernstein (eds) *The Culture Industry: Selected Essays on Mass Culture*, London: Routledge.
Ahearne, Jeremy (2010) *Intellectuals, Culture and Public Policy in France: Approaches from the Left*, Liverpool: Liverpool University Press.
Alberti, Samuel J.M.M. (2009) *Nature and Culture: Objects, Disciplines and the Manchester Museum*, Manchester and New York: University of Manchester Press.
Alexander, Jeffrey (2003) *The Meanings of Social Life: A Cultural Sociology*, Oxford: Oxford University Press.
Alpers, Svetlana (1998) 'The studio, the laboratory, and the vexations of art' in Caroline A. Jones and Peter Galison (eds) *Picturing Science, Producing Art*, London and New York: Routledge.
Anderson, Kay (2007) *Race and the Crisis of Humanism*, London and New York: Routledge.
Anderson, Kay and Colin Perrin (2007) '"The miserablest people in the world": Race, humanism, and the Australian Aborigine', *The Australian Journal of Anthropology*, 18 (1), 18–39.
Anderson, Mark Lynn (2008) 'Taking liberties: the Payne Fund Studies and the creation of the media expert' in Lee Grieveson and Haidee Wasson (eds) *Inventing Film Studies*, Durham, NC and London: Duke University Press.
Archer-Straw, Petrine (2000) *Negrophilia: Avant-Garde Paris and Black Culture in the 1920s*, London: Thames and Hudson.
Arppe, Tiina (2009) 'Sorcerer's apprentices and the "will to figuration". The ambitious heritage of the Collège de Sociologie', *Theory, Culture & Society,* 26 (4), 117–45.
Attwood, Bain (2003) *Rights for Aborigines*, Sydney: Allen and Unwin.
Bagehot, Walter (1873) *Physics and Politics: Or Thoughts on the Application of the Principles of "Natural Selection" and "Inheritance" to Political Society*, London: Henry S. King & Co.
Baker, Keith Michael (1994) 'A Foucauldian French Revolution?' in Jan Goldstein (ed.) *Foucault and the Writing of History*, Oxford: Blackwell.
Baker, Lee D. (2010) *Anthropology and the Racial Politics of Culture*, Durham, NC and London: Duke University Press.
Balzac, Honoré de (2001) *The Unknown Masterpiece*, New York: New York Review of Books.

Barlow, Paul and Colin Trodd (2000) 'Constituting the public: art and its institutions in nineteenth-century London' in Paul Barlow and Colin Trodd (eds) *Governing Cultures: Art Institutions in Victorian London*, Aldershot: Ashgate.

Barnett, Clive (1999) 'Culture, government and spatiality: Reassessing the "Foucault effect" in cultural-policy studies', *International Journal of Cultural Studies*, 2 (3), 369–97.

Barrell, John (1986) *The Political Theory of Painting from Reynolds to Hazlitt*, New Haven, CT and London: Yale University Press.

——(1989) '"The dangerous goddess": masculinity, prestige, and the aesthetic in early eighteenth-century Britain', *Cultural Critique*, spring, 101–31.

——(1992) 'The public prospect and the private view: the politics of taste in eighteenth-century Britain' in *The Birth of Pandora and the Division of Knowledge,* London: Macmillan.

Barthes, Roland (1972) *Mythologies*, London: Jonathan Cape.

Batty, Philip, Lindy Allen and John Morton (eds) (2005) *The Photographs of Baldwin Spencer*, Miegunyah Press: Melbourne.

Belfiore, Eleonara and Oliver Bennett (2010) *The Social Impact of the Arts*, Houndmills: Palgrave Macmillan.

Belting, Hans (2001) *The Invisible Masterpiece*, London: Reaktion Books.

Bennett, Jane (2010) *Vibrant Matter: A Political Ecology of Things*, Durham, NC and London: Duke University Press.

Bennett, Tony (1985) 'Really useless "knowledge": a political critique of aesthetics', *Thesis Eleven*, 12, 28–52.

——(1988) 'The exhibitionary complex', *New Formations*, 4, 73–102.

——(1990) *Outside Literature*, London: Methuen.

——(1992) 'Useful culture', *Cultural Studies*, 6 (3), 395–408.

——(1995) *The Birth of the Museum: History, Theory, Politics*, London and New York: Routledge.

——(1998) *Culture: A Reformer's Science*, Sydney: Allen and Unwin and London: Sage.

——(1999) 'Intellectuals, culture, policy: the technical, the practical and the critical', *Pavis Papers in Social and Cultural Research*, 2, 1–25. Also in Bennett (2007b).

——(2000) 'Acting on the social: art, culture, and government', *American Behavioural Science*, 43 (9), 1412–28. Also in D. Meredyth and J. Minson (eds) *Citizenship and Cultural Policy*, London: Sage Publications, pp. 18–34.

——(2004) *Pasts Beyond Memory: Evolution, Museums, Colonialism*, London and New York: Routledge.

——(2005) 'The historical universal: the role of cultural value in the historical sociology of Pierre Bourdieu', *British Journal of Sociology*, 56 (1), 141–64.

——(2005a) 'Civic laboratories: museums, cultural objecthood, and the governance of the social', *Cultural Studies*, 19 (5), 521–47.

——(2006) 'Exhibition, difference and the logic of culture' in Ivan Karp, Corinne A. Kratz, Lynn Szwaja and Tomás Ybarra-Frausto (eds) *Museum Frictions: Public Cultures/Global Transformations*, Durham, NC: Duke University Press, 46–69.

——(2006a) 'Civic seeing: museums and the organisation of vision', in Sharon MacDonald (ed.) *A Companion to Museum Studies*, Oxford: Blackwell.

——(2007) 'Making culture, changing society: the perspective of culture studies', *Cultural Studies*, 21 (4–5), 610–29.

——(2007a) 'The work of culture', *Journal of Cultural Sociology*, 1 (1), 31–48.

——(2007b) *Critical Trajectories: Culture, Society, Intellectuals*, Malden, MA and Oxford: Blackwell.

——(2007c) 'Habitus clivé: aesthetics and politics in the work of Pierre Bourdieu', *New Literary History*, 38 (1), 201–28.

——(2008/9) 'Aesthetics, government, freedom', *Key Words: A Journal of Cultural Materialism*, 6, 76–91.

——(2009) 'Museum, field, expedition and the circulation of reference,' *Journal of Cultural Economy*, 2 (1–2), 99–116.

——(2010) 'Making and mobilising worlds: assembling and governing the other' in Tony Bennett and Patrick Joyce (eds) *Material Powers: Cultural Studies, History and the Material Turn*, London and New York: Routledge.

——(2010a) 'Culture/knowledge/power: between Foucault and Bourdieu' in Elizabeth Silva and Alan Warde (eds) *Cultural Analysis: The Legacy of Bourdieu*, London and New York: Routledge.

——(2010b) 'Culture studies and the culture complex' in John R. Hall, Laura Grindstaff and Ming-Chen Lo (eds) *Handbook of Cultural Sociology*, London and New York: Routledge.

——(2010c) 'Sociology, aesthetics, expertise', *New Literary History*, 41 (2), 253–76.

——(2011) 'Habit, instinct, survivals: repetition, history, biopolitics' in Simon Gunn and James Vernon (eds) *The Peculiarities of Liberal Modernity in Imperial Britain*, Berkeley, CA: University of California Press.

——(2011a) 'Guided freedom: aesthetics, tutelage and the interpretation of art', *Tate Papers*, 15.

——(2011b) 'Culture, choice, necessity: a political critique of Bourdieu's aesthetic', *Poetics*, 39 (6), 530–46.

——(2012) 'Culture, institution, conduct: the perspective of metaculture' in Meaghan Morris and Metty Hjort (eds) *Instituting Cultural Studies: Creativity and Academic Activism*, Hong Kong: University of Hong Kong Press.

——(2012a) 'Machineries of modernity: museums, theories, histories', *Cultural and Social History*, 9 (1), 145–56.

——(2013) 'The shuffle of things' in Sarah Byrne, Anne Clarke and Rodney Harrison (eds) *Reassembling the Collection: Indigenous Agency and Ethnographic Collections*, Santa Fe, NM: SAR Press.

Bennett, Tony and Chris Healy (eds) (2011) *Assembling Culture*, London: Routledge. (This is a reproduction of a special double issue of the *Journal of Cultural Economy*, 2 [1–2]).

Bennett, Tony and Patrick Joyce (eds) (2010) *Material Powers: Cultural Studies, History and the Material Turn*, London and New York: Routledge.

Bennett, Tony, Mike Savage, Elizabeth Silva, Alan Warde, Modesto Gayo-Cal and David Wright (2009) *Culture, Class, Distinction*, London: Routledge.

Bereson, Ruth (2002) *The Operatic State: Cultural Policy and the Opera House*, London: Routledge.

Blackburn, Kevin (2002) 'Mapping Aboriginal nations: the "nation" concept of late-nineteenth century anthropologists in Australia', *Aboriginal History*, 26, 131–58.

Blanchard, Pascal and Éric Deroo (2008) 'Contrôler: Paris, capitale coloniale (1931–39)' in Pascal Blanchard, Sandrine Lemaire and Nocilas Bancel (eds) *Culture coloniale en France. De la Révolution française à nos jours*, Paris: CNRS Éditions.

Blanchard, Pascal, Sandrine Lemaire and Nocilas Bancel (eds) (2008) *Culture coloniale en France. De la Révolution française à nos jours*, Paris: CNRS Éditions.

Blanckaert, Claude (1988) 'On the origins of French ethnology' in George Stocking Jr. (ed.) *Bones, Bodies, Behaviour: Essays on Biological Anthropology*, Madison, WI: University of Wisconsin Press.

Blanning, Timothy C.W. (2002) *The Culture of Power and the Power of Culture*, Oxford: Oxford University Press.

Boëtsch, Gilles (2008) 'Sciences, savants et colonies (1870–1914)' in Pascal Blanchard, Sandrine Lemaire and Nocilas Bancel (eds) *Culture coloniale en France. De la Révolution française à nos jours*, Paris: CNRS Éditions.

Bohls, Elizabeth A. (1993) 'Disinterestedness and denial of the particular: Locke, Adam Smith, and the subject of aesthetics' in Paul Mattick Jr. (ed.) *Eighteenth Century Aesthetics and the Reconstruction of Art*, Cambridge: Cambridge University Press.

Boltanski, Luc and Ève Chiapello (2007) *The New Spirit of Capitalism*, London and New York: Verso.

Bolton, Lissant (2003) *'Living and Dying:* ethnography, class and aesthetics at the British Museum', paper presented at the *Museums and Difference* conference, Centre for 21st Century Studies, University of Wisconsin-Milwaukee and the Milwaukee Art Museum, November.

Bondas, Julien (2011) 'L'ethnographie comme chasse: Michel Leiris et les animaux de la mission Dakar-Djibouti', *Gradhiva*, 13, 162–81.

Borsay, Peter (1989) *The English Urban Renaissance: Culture and Society in the English Provincial Town, 1660–1770*, Oxford: Clarendon Press.

Bourdieu, Pierrre (1977) *Outline of a Theory of Practice*, Cambridge: Cambridge University Press.

——(1979) 'Public opinion does not exist' in Armand Mattelart and Seth Siegelaub (eds) *Communication and Class Struggle*, Vol. 1, New York/Bagnolet: International General/IMMRC.

——(1980) 'Le mort saisit le vif: les relations entre l'histoire réifiée et l'histoire incorporée', *Actes de la recherche en sciences sociae,* 32–3, 3–14.

——(1984) *Distinction: A Social Critique of the Judgement of Taste*, London and New York: Routledge.

——(1987) 'What makes a social class? On the theoretical and practical existence of groups', *Berkeley Journal of Sociology*, 32, 1–17.

——(1988) *Homo Academicus*, Cambridge: Polity.

——(1990) *In Other Words*, Cambridge: Polity Press.

——(1996) *The Rules of Art: Genesis and Structure of the Literary Field*, Cambridge: Polity Press.

——(1998) *Practical Reason: On the Theory of Action*, Cambridge: Polity.

——(2000) *Pascalian Meditations*, Cambridge: Polity.

——(2001) *Masculine Society*, Cambridge: Polity.

——(2004) *Science of Science and Reflexivity*, Chicago: University of Chicago Press.

——(2005) *The Social Structures of the Economy*, Cambridge: Polity.

Bourdieu, Pierre and Alain Darbel (1991) *The Love of Art: European Art Museums and Their Public*, Cambridge: Polity.

Bourdieu, Pierre and Loïc Wacquant (1992) *An Invitation to Reflexive Sociology*, Chicago: University of Chicago Press.

Bowland, Tom (2007) 'Critique as a technique of self: a Butlerian analysis of Judith Butler's prefaces', *History of the Human Sciences*, 20 (3), 105–22.

Briggs, Charles and Richard Bauman (1999) '"The foundation of all future researches": Franz Boas, George Hunt, Native American texts, and the construction of modernity', *American Quarterly*, 51 (3), 479–528.

Brown, Bill (2003) *A Sense of Things: The Object Matter of American Literature*, Chicago, IL and London: University of Chicago Press.

Brown, Steven D. (2002) 'Michel Serres: science, translation and the logic of the parasite', *Theory, Culture & Society*, 19 (3), 1–27.

Brown, Vivienne (1994) *Adam Smith's Discourse: Canonicity, Commerce and Conscience*, London and New York: Routledge.

——(1997) 'Dialogism, the gaze, and the emergence of economic discourse', *New Literary History*, 28, 697–710.

Brown, Wendy (2006) *Regulating Aversion: Tolerance in the Age of Identity and Empire*, Princeton, NJ: Princeton University Press.

Bunzl, Matti (1996) 'Franz Boas and the Humboldtian tradition: from *Volkgeist* and *Nationalcharakter* to an anthropological concept of culture' in George W. Stocking Jr. (ed.) *Volkgeist as Method: Essays on Boasian Ethnography and the German Anthropological Tradition*, Madison, WI: University of Wisconsin Press.

Caliskan, Koray and Michel Callon (2009) 'Economisation, part 1: shifting attention from the economy towards processes of economisation', *Economy and Society*, 38 (3), 369–398.

——(2010) 'Economisation, part 2: a research programme for the study of markets', *Economy and Society*, 39 (1), 1–32.

Callewaert, Stefan (2006) 'Bourdieu, critic of Foucault: the case against double-game-philosophy', *Theory, Culture & Society*, 23 (6), 73–98.

Callon, Michel (2005) 'Why virtualism paves the way to political impotence: a reply to Daniel Miller's critique of *The Laws of the Markets*', *Economic Sociology: European Electronic Newsletter*, February, 6 (2), 3–20.

Callon, Michel and Bruno Latour (1981) 'Unscrewing the big Leviathan: how actors macro-structure reality and how sociologists help them to do so' in K. Knorr-Cetina and A. Cicourel (eds) *Advances in Social Theory and Methodology*, London: Routledge.

Callon, Michel and John Law (1995) 'Agency and the hybrid *collectif*', *South Atlantic Quarterly*, 94, 481–507.

Callon, Michel, Yuval Millo and Fabian Muniesa (eds) (2007) *Market Devices*, Oxford: Blackwell/The Sociological Review.

Campbell, Colin (1996) *Detraditionalisation: Critical Reflections on Authority and Identity*, Oxford: Blackwell.

Candea, Matei (ed.) (2010) *The Social after Gabriel Tarde: Debates and Assessments*, London: Routledge.

Canguilhem, Georges (2008) *Knowledge of Life*, New York: Fordham University Press.

Cantrill, Arthur and Corinne Cantrill (1982) 'The 1901 cinematography of Baldwin Spencer', *Cantrill's Filmnotes,* 37/38, 27–42.

Carey, John (2005) *What Good are the Arts?* London: Faber and Faber.

Carroll, Patrick (2006) *Science, Culture, and Modern State Formation*, Berkeley, Los Angeles and London: University of California Press.

Caygill, Howard (1989) *Art of Judgement*, Oxford: Basil Blackwell.

Chambers, Samuel (2011) 'The politics of the police: from neoliberalism to anarchism and back to democracy' in Paul Bowman and Richard Stamp (eds) *Reading Rancière*, New York: Continuum.

Chytry, Josef (1989) *The Aesthetic State: A Quest in Modern German Thought*, Berkeley, Los Angeles and London: University of California Press.

Clarke, John (2009) 'Programmatic statements and dull empiricism: Foucault's neo-liberalism and social policies', *Journal of Cultural Economy*, 2 (1–2), 229–33.

Clarke, Tom and Galligan, Brian (1995) 'Aboriginal native and the institutional construction of the Australian 1901–48', *Australian Historical Studies* 26 (105): 523–43.

Clifford, James (1983) 'Power and dialogue in ethnography: Marcel Griaule's initiation' in George W. Stocking (ed.) *Observers Observed: Essays on Ethnographic Fieldwork*, Madison: University of Wisconsin Press.

——(1988) *The Predicament of Culture: Twentieth-Century Ethnography, Literature, and Art*, Cambridge, MA: Harvard University Press.

——(2007) 'Quai Branly in process', *October*, 120, 2–23.

Collier, Stephen J. (2009) 'Topologies of power: Foucault's analysis of political government beyond governmentality', *Theory, Culture & Society*, 26 (6), 78–108.

Collini, Stefan (1979) *Liberalism and Sociology: L.T. Hobhouse and Political Argument in England, 1880–1914*, Cambridge: Cambridge University Press.

——(1991) *Public Moralists: Political Thought and Intellectual Life in Britain, 1850–1930*, Oxford: Clarendon Press.

Commonwealth Government, Human Rights and Equal Opportunity Commission (1997) *Bringing them Home: National Inquiry into the Separation of Aboriginal and Torres Strait Children from their Families*, Canberra: Australian Government Publishing Service.

Conklin, Alice C. (1997) *A Mission to Civilise: The Republican Ideal of Empire in France and West Africa*, New York: Peter Lang.

——(2002) 'The new "ethnology" and "la situation colonial" in interwar France', *French Politics, Culture & Society*, 20 (2), 29–46.

——(2002a) 'Civil society, science, and empire in late republican France: The foundation of Paris's Museum of Man', *Osiris*, 17, 255–90.

——(2008) 'Skulls on display: the science of race in Paris's Musée de l'Homme, 1928–50' in Daniel Sherman (ed.) *Museums and Difference*, Bloomington and Indianapolis: Indiana University Press.

Connell, Philip (2001) *Romanticism, Economics and the Question of 'Culture'*, Oxford: Oxford University Press.

Corcuff, Philippe (2001) 'Le collectif au défi du singulier: en partant de l'habitus' in Bernard Lahire (ed.) *Le travail sociologique de Pierre Bourdieu: Dettes et critique*, Paris: Éditions La Découverte.

Crary, Johnathan (2001) *Suspensions of Perception: Attention, Perception, and Modern Culture*, Cambridge, MA: MIT Press.

Crook, Tom (2007) 'Power, privacy and pleasure: liberalism and the modern cubicle', *Cultural Studies*, 21 (4–5), 549–69.

Curtis, Bruce (2002) 'Foucault on governmentality and population: the impossible discovery', *Canadian Journal of Sociology*, 27 (4), 505–33.

Darwin, Charles (1881) *The Formation of Vegetable Mould through the Action of Worms with Observations on their Habits*, London: John Murray.

——(1952) *The Descent of Man, and Selection in Relation to Sex*, Chicago, IL: William Benton.

Daston, Lorraine and Peter Galison (2010) *Objectivity*, New York: Zone Books.

Dean, Jodi (2011) 'Politics without police' in Paul Bowman and Richard Stamp (eds) *Reading Rancière*, New York: Continuum.

Dean, Mitchell (2007) *Governing Societies: Political Perspectives on Domestic and International Rule*, Maidenhead: Open University Press.

Debaene, Vincent (2006) '"Étudier des états de conscience". La reinvention du terrain par l'ethnologie, 1925–39', *L'Homme*, 179, 7–62.

——(2010) *L'adieu au voyage: L'ethnologie française entre science et littérature*, Paris: Éditions Gallimard.

Defert, Daniel (1982) 'A collection of the world: accounts of voyages from the sixteenth to the eighteenth centuries', *Dialectical Anthropology*, 7 (1), 11–20.

DeLanda, Manuel (2006) *A New Philosophy of Society*, London and New York: Continuum.

Deleuze, Gilles (1984) *Kant's Critical Philosophy: The Doctrine of the Faculties*, Minneapolis, MN: University of Minnesota Press.

——(1999) *Foucault*, London: The Athlone Press.

Deleuze, Gilles and Felix Guattari (1988) *A Thousand Plateaus: Capitalism and Schizophrenia*, London: The Athlone Press.

Deleuze, Gilles and Claire Parnet (2002) *Dialogues II*, New York: Columbia University Press.

Derrida, Jacques (1998) *Archive Fever: A Freudian Impression*, Chicago, IL: University of Chicago Press.

Desvallées, André (2007) *Quai Branly: Un Miroir aux Alouettes? A propos d'ethnographie et d'arts premiers'*, Paris: L'Harmattan.

Dias, Nélia (1991) *Le Musée d'Ethnographie du Trocadéro (1878–1908). Anthropologie et muséologie en France*, Paris: Éditions du Centre National de la Recherche Scientifique.

——(2004) *La mesure des sens: Les anthropologues et le corps humain aux XIXe siècle*, Paris: Aubier.

——(2006) '"What's in a name?" Anthropology, museums and values' in Cordula Crewe (ed.) *Die Schau des Fremden: Ausstellungskonzepte zwischen Kunst, Kommerz und Wissenschaft (Transatlantische Historische Studien, Bd 26)*, Stuttgart: Franz Steiner Verlag.

——(2010) 'Exploring the senses and exploiting the land: railroads, bodies and measurement in nineteenth-century French colonies' in Tony Bennett and Patrick Joyce (eds) *Material Powers: Cultural Studies, History and the Material Turn*, London and New York: Routledge.

Didi-Huberman, Georges (2002) 'The surviving image: Aby Warburg and Taylorian anthropology', *Oxford Art Journal*, 25 (1), 59–70.

Dillon, Michael (2004) 'The security of governance' in Wendy Larner and William Walters (eds) *Global Governmenality: Governing International Spaces*, London: Routledge.

Dowling, Linda (1996) *The Vulgarisation of Art: The Victorians and Aesthetic Democracy*, Charlottesville, VA and London: University Press of Virginia.

du Gay, Paul (2000) *In Praise of Bureaucracy: Weber, Organisation, Ethics*, London: Sage.

Dupont, Danica and Frank Pearce (2001) 'Foucault contra Foucault: rereading the "governmentality" papers', *Theoretical Criminology*, 5 (2), 123–58.

During, Simon (2010) *Exit Capitalism: Literary Culture, Theory, and Post-Secular Modernity*, London and New York: Routledge.

Durkheim, Emile (1964) *The Division of Labour in Society*, New York: The Free Press and London: Collier-Macmillan.

——(1968) *The Elementary Forms of the Religious Life*, London: George Allen and Unwin.

Eagleton, Terry (1990) *The Ideology of the Aesthetic*, Oxford: Blackwell.

Elias, Norbert (1978) *The History of Manners*, New York: Pantheon Books.

Fabian, Johannes (1983) *Time and the Other: How Anthropology Makes Its Object*, New York: Columbia University Press.

——(2000) *Out of Our Minds: Reason and Madness in the Exploration of Central Africa*, Berkeley, Los Angeles and London: University of California Press.

Fabre, Daniel (1997) 'L'ethnologie française a la croisée des engagements (1940–45)' in Jean-Yves Boursier (ed.) *Resistants et Resistance*, Paris: Éditions L'Harmattan.

Fisher, Philip (1996) 'Local meanings and portable objects: national collections, literatures music, and architecture' in Gwendolyn Wright (ed.) *The Formation of National Collections of Art and Archaeology*, Washington, DC: National Gallery of Art.

Foucault, Michel (1973) *The Birth of the Clinic: An Archaeology of Medical Perception*, London: Tavistock Publications.

——(1988) *The Care of the Self*, New York: Vintage Books.

——(1989) 'The aesthetics of existence' in Sylvère Lotringer (ed.) *Foucault Live (Interviews, 1966–84)*, New York: Semiotext(e).

——(1991) 'Governmentality' in Graham Burchell, Colin Gordon and Peter Miller (eds) *The Foucault Effect: Studies in Governmentality*, London: Harvester Wheatsheaf.

——(2003) *Society Must Be Defended: Lectures at the Collège de France, 1975–76*, New York: Picador.

——(2005) *The Hermeneutics of the Subject; Lectures at the Collège de France, 1981–82*, New York: Picador.

——(2007) *Security, Territory, Population: Lectures at the Collège de France, 1977–1978*, London: Palgrave Macmillan.

——(2007a) *The Politics of Truth*, Los Angeles, CA: Semiotext(e).

——(2008) *The Birth of Biopolitics: Lectures at the Collège de France, 1978–79*, London: Palgrave Macmillan.

——(2010) *The Government of Self and Others: Lectures at the Collège de France, 1982–1983*, London: Palgrave Macmillan.

Fournier, Marcel (2006) *Marcel Mauss: A Biography*, Princeton, NJ and Oxford: Princeton University Press.

Frank, Johann Peter (1976) *A System of Complete Medical Police*, ed. Erna Lensky, Baltimore, MD: Johns Hopkins University Press.

Frow, John (2004) 'A pebble, a camera, a man who turns into a telegraph pole' in Bill Brown (ed.) *Things*, Chicago, IL and London: University of Chicago Press.

Gell, Alfred (1998) *Art and Agency: An Anthropological Theory*, Oxford: Clarendon Press.

Goodlad, Lauren M.E. (2003) *Victorian Literature and the Victorian State: Character and Governance in a Liberal Society*, Baltimore, MD and London: The Johns Hopkins University Press.

Goodman, Jane E. and Paul A. Silverstein (eds) (2009) *Bourdieu in Algeria: Colonial Politics, Ethnographic Practices, Theoretical Developments*, Lincoln, NE and London: University of Nebraska Press.

Gorgus, Nina (2003) *Le magician des vitrines: Le muséologue Georges Henri Rivière*, Paris: Éditions de la Maison des sciences de l'homme.

Gosden, Chris and Frances Larson, with Alison Petch (2007) *Knowing Things: Exploring the Collections at the Pitt Rivers Museum, 1884–1945*, Oxford: Oxford University Press.

Goswami, Manu (2004) *Producing India: From Colonial Economy to National Space*, Chicago, IL: University of Chicago Press.

Griaule, Marcel (1933) 'Introduction méthodologique', *Minotaure*, 2, 7–12.

——(1991) *Les flambeurs d'homme*, Paris: Berg International.

Grieveson, Lee (2009) 'On governmentality and screens', *Screen*, 50 (1), 180–87.

Griffiths, Tom (1996) *Hunters and Collectors: The Antiquarian Imagination in Australia*, Cambridge: Cambridge University Press.

Grognet, Fabrice (2010) *Le Concept de Musée. La patrimonialisation de la culture des 'autres'. D'une rive a l'autre, du Trocadéro a Branly: histoire de metamorphoses*, Thèse de doctorat en Ethnologie, École des Hautes Études en Sciences Sociales.

Grossberg, Lawrence (2010) *Cultural Studies in the Future Tense*, Durham, NC and London: Duke University Press.

Guyer, Paul (2005) *Values of Beauty: Historical Essays in Aesthetics*, Cambridge: Cambridge University Press.

Habermas, Jürgen (1989) *The Structural Transformation of the Public Sphere – An Inquiry into a Category of Bourgeois Society*, Cambridge: Polity.

Hacking, Ian (1992) *Representing and Intervening*, Cambridge: Cambridge University Press.

——(1992a) 'The self-vindication of the laboratory sciences' in Andrew Pickering (ed.) *Science as Practice and Culture*, Chicago, IL and London: University of Chicago Press.

Hadot, Pierre (1995) *Philosophy as a Way of Life*, Oxford: Blackwell.

Haebich, Anna (2000) *Broken Circles: Fragmenting Indigenous Families, 1800–2000*, Fremantle: Fremantle Arts Centre Press.

Hage, Ghassan (2008) 'Analysing multiculturalism today' in Tony Bennett and John Frow (eds) *The SAGE Handbook of Cultural Analysis*, London: SAGE.

Hall, Martin (2006) 'The reappearance of the authentic' in Ivan Karp, Corinne A. Kratz, Lynn Szwaja and Tomás Ybarra-Frausto (eds) *Museum Frictions: Public Cultures/ Global Transformations*, Durham, NC: Duke University Press.

Hanway, Jonas (1775) *The Defects of Police, the Causes of Immorality*, London: J. Dodsley.

Haraway, Donna (1997) *Modest_Witness@Second_Millennium. FemaleMan©_Meets_ OncoMouse^{TM}*, New York and London: Routledge.

Harman, Graham (2009) *Prince of Networks: Bruno Latour and Metaphysics*, Melbourne: re.press.

Hawkins, Mike (1997) *Social Darwinism in European and American Thought, 1860–1945*, Cambridge: Cambridge University Press.

Hay, James and Mark Andrejevic (2006) 'Homeland insecurities', *Cultural Studies*, 20 (4–5), 331–48.

Healy, Chris (2008) *Forgetting Aborigines*, Sydney: University of New South Press.

Hecht, Jennifer Michael (2003) *The End of the Soul: Scientific Modernity, Atheism, and Anthropology in France*, New York: Columbia University Press.

Héran, François (1987) 'La second nature de l'habitus. Tradition philosophique et sens commun dans le langage sociologique', *Revue française de sociologie*, 28 (3), 385–416.

Herder, Johann Gottfried (2002) *Sculpture: Some Observations on Shape and Form from Pygmalion's Creative Dream*, Chicago, IL and London: University of Chicago Press.

Herle, Anita and Sandra Rouse (eds) (1998) *Cambridge and the Torres Strait: Centenary Essays on the 1898 Anthropological Expedition*, Cambridge: Cambridge University Press.

Hetherington, Kevin (2007) *Capitalism's Eye: Cultural Spaces and the Commodity*, London: Routledge.

——(2011) 'Foucault, the museum and the diagram', *Sociological Review*, 59 (3), 457–75.

Hill, Barry (2002) *Broken Song: F.G.H. Strehlow and Aboriginal Possession*, Sydney and New York: Vintage.

Hinsley, Curtis M., Jr. (1981) *Savages and Scientists: The Smithsonian Institution and the Development of American Anthropology 1846–1910*, Washington, DC: Smithsonian Institution Press.

Hogarth, William (1997 [1753]) *The Analysis of Beauty*, New Haven, CT and London: Yale University Press for the Paul Mellon Centre for British Art.

Hoggart, R. (1969) *The Uses of Literacy*, Harmondsworth: Penguin.

Hollier, Denis and Liesl Ollman (1991) 'The use-value of the impossible', *October*, 60, 3–24.

Hooper-Greenhill, Eilean (1992) *Museums and the Shaping of Knowledge*, London and New York: Routledge.

Hunter, Ian (1988) *Culture and Government: The Emergence of Literary Education*, London: Macmillan.

——(1991) 'Personality as a vocation: the political rationality of the humanities', *Economy and Society*, 19 (4), 391–430.

——(2001) *Rival Enlightenments: Civil and Metaphysical Philosophy in Early Modern Germany*, Cambridge: Cambridge University Press.

Jacknis, Ira (1996) 'The ethnographic object and the object of ethnology in the early career of Franz Boas' in George Stocking, Jr. (ed.) *Volksgeist as Method and Ethic: Essays on Boasian Ethnography and the German Anthropological Tradition*, Madison, WI: University of Wisconsin Press.

Jamin, Jean (1982) 'Objets trouvé des paradis perdus. A propos de la Mission Dakar-Djibouti' in Jacques Hainard and Roland Kachr (eds) *Collections passions*, Neuchâtel: Musée d'ethnographie.

——(1986) 'D'ethnographie mode d'inemploi. De quelques rapports de l'ethnologie avec la malaise dans la civilisation' in Jacques Hainard and Roland Kaehr (eds) *Le mal et le douleur*, Neuchâtel: Musée d'ethnographie.

——(1998) 'Tout était fétiche, tout devint totem' preface to Jean Jamin (ed.) *Bulletin du Musée d'Ethnographie du Trocadero*, Paris: Jean Michel Place.

Jolly, Éric (2001) 'Marcel Griaule, ethnologue: La construction d'une discipline (1925–56)', *Journal des africanistes*, 71 (1), 149–90.

Jones, Philip (1996) 'The Horne Expedition's place among nineteenth-century inland exhibitions' in S. R. Morton and Derek John Mulvaney (eds) *Exploring Central Australia: Society, Environment and the 1894 Expedition*, Chipping Nortin, NSW: Surrey Beatty.

——(2007) *Ochre and Rust: Artefacts and Encounters on Australian Frontiers*, Kent Town, SA: Wakefield Press.

Joyce, Patrick (2002) *The Social in Question. New Bearings in History and the Social Sciences*, London: Routledge.

——(2003) *The Rule of Freedom: Liberalism and the Modern City*, London: Verso.

Kalpagam, U. (2002) 'Colonial governmentality and the public sphere in India', *Journal of Historical Sociology*, 15 (1), 35–58.

——(2002a) 'Colonial governmentality and the "economy"', *Economy and Society*, 29 (3), 418–38.

Kant, Immanuel (1987) *Critique of Judgement*, Indianapolis, IN and Cambridge: Hackett Publishing Company.

——(2006 [1798]) *Anthropology from a Pragmatic Point of View*, trans. Robert L. Louden, Cambridge: Cambridge University Press.

Klein, Lawrence E. (1994) *Shaftesbury and the Culture of Politeness: Moral Discourse and Cultural Politics in Early Eighteenth-Century England*, Cambridge: Cambridge University Press.

Knorr-Cetina, Karin (1992) 'The couch, the cathedral, and the laboratory: on the relationship between experiment and laboratory in science' in Andrew Pickering (ed.) *Science as Practice and Culture*, Chicago, IL and London: University of Chicago Press.

——(1997) 'Sociality with objects: social relations in postsocial knowledge societies', *Theory, Culture and Society*, 14 (4), 1–28.

——(1999) *Epistemic Cultures: How the Sciences Make Knowledge*, Cambridge, MA: Harvard University Press.

——(2001) 'Objectual practice' in Theodore R. Schatzki, Karin Knorr-Cetina and Eike von Savigny (eds) *The Practice Turn in Contemporary Theory*, London and New York: Routledge.

Knox, Hannah, Mike Savage and Penny Harvey (2005) 'Social networks and spatial relations: networks as method, metaphor and form', *CRESC Working Papers on Socio-cultural Change*, no. 1, ESRC Centre for Research on Socio-cultural Change, University of Manchester and The Open University.

Koselleck, Reinhart (2002) *The Practice of Conceptual History: Timing History, Spacing Concepts*, Stanford, CA: Stanford University Press.

Kuklick, Henrika (2006) '"Humanity in the chrysalis stage": indigenous Australians in the anthropological imagination, 1899–1926', *British Journal for the History of Science*, 39 (4), 535–68.

——(2011) 'Reflections on the history of fieldwork, with special reference to sociocultural anthropology', *Isis*, 102, 1–38.

Kuper, Adam (1999) *Culture: The Anthropologists' Account*, Cambridge, MA: Harvard University Press.

Latour, Bruno (1987) *Science in Action: How to Follow Scientists and Engineers through Society*, Cambridge, MA: Harvard University Press.

——(1988) *The Pasteurisation of France*, Cambridge, MA: Harvard University Press.

——(1990) 'Drawing things together' in Michael Lynch and Steve Woolgar (eds) *Representation in Scientific Practice*, Cambridge, MA: MIT Press.

——(1993) *We Have Never Been Modern*, Cambridge, MA: Harvard University Press.

——(1998) 'How to be iconophilic in art, science and religion?' in Caroline A. Jones and Peter Galison (eds) *Picturing Science, Producing Art*, London and New York: Routledge.

——(1999) 'Give me a laboratory and I will raise the world' in Mario Biagioli (ed.) *The Science Studies Reader*, New York and London: Routledge.

——(1999a) *Pandora's Hope: Essays on the Reality of Science Studies*, Cambridge, MA: Harvard University Press.

——(2002) 'Morality and technology: the end of the means', *Theory, Culture & Society*, 19 (5/6), 247–60.

——(2002a) 'Gabriel Tarde and the end of the social' in Patrick Joyce (ed.) *The Social in Question: New Bearings in History and the Social Sciences*, London and New York: Routledge.

——(2004) 'Why has critique run out of steam: from matters of fact to matters of concern' in Bill Brown (ed.) *Things*, Chicago, IL: University of Chicago Press.

——(2004a) *Politics of Nature: How to Bring the Sciences Into Democracy*, Cambridge, MA: Harvard University Press.

——(2005) *Reassembling the Social: An Introduction to Actor-Network Theory*, Oxford: Oxford University Press.

——(2005a) 'From realpolitik to dingpolitik or how to make things public' in Bruno Latour and Peter Weibel (eds) *Making Things Public: Atmospheres of Democracy*, Karlsruhe: ZKM/Centre for Art and Cambridge, MA: MIT Press.

——(2010) 'An attempt at a "compositionist manifesto"', *New Literary History*, 41 (3), 471–90.

Latour, Bruno and Steve Woolgar (1986) *Laboratory Life: The Construction of Scientific Facts*, Princeton, NJ: Princeton University Press.

Laurière, Christine (2008) *Paul Rivet le savant et le politique*, Paris: Muséum national d'Histoire naturelle.

Law, John (2004) *After Method: Mess in Social Science Research*, London and New York: Routledge.

Law, John and John Urry (2004) 'Enacting the social', *Economy and Society*, 33 (3), 390–410.

Lawrence, Rebecca and Chris Gibson (2007) 'Obliging Indigenous citizens? Shared Responsibility Agreements in Australian Aboriginal communities', *Cultural Studies*, 21 (4–5), 650–71.

Lebovics, Herman (1992) *True France: The Wars over Cultural Identity, 1900–1945*, Ithaca, NY and London: Cornell University Press.

——(2004) *Bringing the Empire Back Home: France in the Global Age*, Durham, NC and London: Duke University Press

Lee, Richard E. (2003) *Life and Times of Cultural Studies: The Politics and Transformation of the Structures of Knowledge*, London: Lawrence and Wishart.

Leiris, Michel (1996) 'L'Afrique fantôme' in *Miroir de l'Afrique*, Paris: Éditions Gallimard.

Legg, Stephen (2007) *Spaces of Colonialism: Delhi's Urban Governmentalities*, Oxford: Blackwell.

L'Estoile, Benoît de (2007) *Le Goût des Autres: De l'Exposition coloniale aux Arts premiers*, Paris: Flammarion.

Lewis, Geoffrey (1989) *For Instruction and Recreation – A Century History of the Museums Association*, London: Quiller Press.

Li, Tania Murray (2007) 'Practices of assemblage and community forest management', *Economy and Society*, 36 (2), 263–93.

Long, Paul (2008) *Only in the Common People: The Aesthetics of Class in Post-War Britain*, Newcastle: Cambridge Scholars Publishing.

Luhmann, Niklas (2000) *Art as a Social System*, Stanford, CA: Stanford University Press.

Lukács, Georg (1971) *History and Class Consciousness*, London: Merlin Press.

Lumley, Robert (ed.) (1988) *The Museum Time-Machine: Putting Cultures on Display*, London and New York: Comedia/Methuen.

Lydon, Jane (2005) *Eye Contact: Photographing Indigenous Australians*, Durham, NC and London: Duke University Press.

MacGregor, Neil (2003) 'Preface' in Kim Sloan (ed.) *Enlightenment: Discovering the World in the Eighteenth Century*, London: The British Museum Press.

MacKenzie, John (2009) *Museums and Empire: Natural History, Human Cultures and Colonial Identities*, Manchester: Manchester University Press.

Malabou, Catherine (2008) 'Addiction and grace' in Félix Ravaisson, *Of Habit*, London and New York: Continuum.

Maleuvre, Didier (1999) *Museum Memories: History, Technology, Art*, Stanford, CA: Stanford University Press.

Maltz, Diana (2006) *British Aestheticism and the Urban Working Classes, 1870–1900. Beauty for the People*, Houndmills: Palgrave Macmillan.

Maudsley, Henry (1902) *Life in Mind and Conduct*, London: Macmillan & Co.

McCarthy, Conal (2007) *Exhibiting Maori: A History of Colonial Cultures of Display*, Oxford and New York: Berg.

McClellan, Andrew (2003) 'A brief history of the art museum public' in Andrew McClellan (ed.) *Art and its Publics: Museum Studies at the Millennium*, Oxford: Blackwell.

McDonald, Rónán (2008) 'A culture of excellence', *The Guardian*, 12 January, 39.

McFall, Liz (2009) 'The agencement of industrial branch life assurance', *Journal of Cultural Economy*, 2 (1–2), 49–66.

McGregor, Russell (1997) *Imagined Destinies: Aboriginal Australians and the Doomed Race Theory, 1880–1939*, Melbourne: Melbourne University Press.

——(2002) '"Breed out the colour", or the importance of being white', *Australian Historical Studies*, 120, 286–302.

McMaster, Sir Brian (2008) *Supporting Excellence in the Arts*, London: Department of Culture, Media and Sport.

McNeill, Donald (2010) 'Behind the "Heathrow hassle": a political and cultural economy of the privatized airport', *Environment and Planning*, 42 (12), 2859–73.

Marin, Louis (1988) *Portrait of the King*, Basingstoke: Macmillan.

Mauss, Marcel (1913 [1969]) 'L'ethnographie en France et à l'étranger' in *Ouevres, v3: Cohésion social et divisions de la sociologie*, Paris: Les Éditions de Minuit.

——(1920) 'L'état actuel des sciences anthropologiques en France', *Ouevres, v. 3: Cohésion social et divisions de la sociologie*, Paris: Les Éditions de Minuit.

——(2007) *Manual of Ethnography*, New York and Oxford: Durkheim Press/Berghahn Books.

Mehta, Uday S. (1997) 'Liberal strategies of exclusion' in Frederick Cooper and Ann Laura Stoler (eds) *Tensions of Empire: Colonial Cultures in a Bourgeois World*, Berkeley, Los Angeles and London: University of California Press.

Métraux, Alfred (1941) *L'île de Pâques*, Paris: Éditions Gallimard.

Mill, John Stuart (1967 [1843]) *A System of Logic, Ratiocinative and Deductive. Being a Connected View of the Principles of Evidence and the Methods of Scientific Investigation*, London: Longmans.

——(1969) *On Liberty; Representative Government; The Subjection of Women: Three Essays*, London: Oxford University Press.

Mitchell, Timothy (2002) *Rule of Experts: Egypt, Techno-Politics, Modernity*, Berkeley and Los Angeles: University of California Press.

Mol, Annemarie (2002) *The Body Multiple: Ontology in Medical Practice*, Durham, NC: Duke University Press.

Morgan, Lloyd (1896) *Habit and Instinct*, London: Edward Arnold.

Morphy, Howard (1996) 'Empiricism to metaphysics: in defence of the concept of the Dreamtime' in Tim Bonyhady and Tom Griffiths (eds) *Prehistory to Politics: John Mulvaney, the Humanities and the Public Intellectual*, Melbourne: Melbourne University Press.

——(1996a) 'More than mere facts: repositioning Spencer and Gillen in the history of anthropology' in S. R. Morton and Derek John Mulvaney (eds) *Exploring Central Australia: Society, Environment and the 1894 Expedition*, Chipping Nortin, NSW: Surrey Beatty.

——(2007) *Becoming Art: Exploring Cross-Cultural Categories*, Oxford and New York: Berg.

Morris, Meaghan (2006) *Identity Anecdotes: Translation and Media Culture*, London: Sage.

Mukerji, Chandra (1997) *Territorial Ambitions and the Gardens of Versailles*, Cambridge: Cambridge University Press.

——(2010) 'The intended state' in Tony Bennett and Patrick Joyce (eds) *Material Powers: Cultural Studies, History and the Material Turn*, London and New York: Routledge.

Mulhern, Francis (2000) *Culture/Metaculture*, London and New York: Routledge.

Mulvaney, D. J. and J. H. Calaby (1985) *'So Much That is New': Baldwin Spencer, 1860–1929: A Biography*, Melbourne: Melbourne University Press.

Muthu, Sankar (2003) *Enlightenment against Empire*, Princeton, NJ: Princeton University Press.

Myers, Fred (2002) *Painting Culture: The Making of an Aboriginal High Art*, Durham, NC: Duke University Press.

Nakata, Martin (2007) *Disciplining the Savages, Savaging the Disciplines*, Canberra: Aboriginal Studies Press.

Nimmo, Richie (2010) *Milk, Modernity and the Making of the Human*, London and New York: Routledge.

O'Connor, Justin (2010) *Arts and the Creative Industries: A Report for the Australia Council*, Sydney: Australia Council.

Ogborn, Miles (2007) *Indian Ink: Script and Print in the Making of the English East India Company*, Chicago, IL and London: University of Chicago Press.

O'Malley, Pat (1998) 'Indigenous governance' in Mitchell Dean and Barry Hindess (eds) *Governing Australia: Studies in Contemporary Rationalities of Government*, Cambridge: Cambridge University Press.

Osborne, Peter (ed.) (2000) *From an Aesthetic Point of View: Philosophy, Art and the Senses*, London: Serpent's Tail.

——(2006) '"Whoever speaks of culture speaks of administration as well": disputing pragmatism in cultural studies', *Cultural Studies*, 20 (1), 33–47.

Osborne, Thomas (1998) *Aspects of Enlightenment: Social Theory and the Ethics of Truth*, London: UCL Press.

——(2008) *The Structure of Modern Cultural Theory*, Manchester and New York: Manchester University Press.

Osborne, Thomas and Nikolas Rose (1999) 'Do the social sciences create phenomena: the case of public opinion research', *British Journal of Sociology*, 50 (3), 367–96.

——(2008) 'Populating sociology: Carr-Saunders and the problem of population', *Sociological Review*, 56 (4), 552–78.

Otis, Laura (1994) *Organic Memory: History and Body in the Late Nineteenth and Early Twentieth Centuries*, Lincoln, NE and London: University of Nebraska Press.

Otter, Chris (2007) 'Making liberal objects: British techno-social relations 1800–1900', *Cultural Studies*, 21 (4–5), 570–90.

——(2008) *The Victorian Eye: A Political History of Light and Vision in Britain, 1800–1910*, Chicago, IL: University of Chicago Press.

Ouellette, Laurie (2002) *Viewers Like You? How Public TV Failed the People*, New York: Columbia University Press.

Ouellette, Laurie and James Hay (2008) *Better Living through Reality TV*, Oxford: Blackwell.

Pagden, Anthony (1986) *The Fall of Natural Man: The American Indian and the Origins of Comparative Ethnology*, Cambridge: Cambridge University Press.

Patton, Paul (2000) *Deleuze and the Political*, London and New York: Routledge.

Pearson, Noel (2011) 'Constitutional reform crucial to indigenous wellbeing', *The Weekend Australian*, 24–25 December, 20.

Peer, Shanny (1998) *France on Display: Peasants, Provincials, and Folklore in the 1937 Paris World's Fair*, Albany, NY: SUNY Press.

Penny, Glenn H. (2002) *Objects of Culture: Ethnology and Ethnographic Museums in Imperial Germany*, Chapel Hill, NC and London: University of North Carolina Press.

Pickering, Andrew (2001) 'Practice and posthumanism: social theory and a history of agency' in Theodore R. Schatzki, Karin Knorr Cetina and Eike von Savigny (eds) *The Practice Turn in Contemporary Social Theory*, London and New York: Routledge.

Pickstone, John (1994) 'Museological science', *History of Science*, 32 (2), 111–38.

——(2000) *Ways of Knowing: A New History of Science, Technology and Medicine*, Manchester: Manchester University Press.

Pierre, Anne Laure (2001/2) 'Ethnographie et photographie. La mission Dakar-Djibouti', *Gradhiva*, 30/31, 105–13.

Pitt Rivers, Henry (published under Col. A. H. Lane Fox) (1875) 'On the principles of classification adopted in the arrangement of his anthropological collection now exhibited in the Bethnal Green Museum', *Journal of the Anthropological Institute of Great Britain and Ireland*, 4.

Poovey, Mary (1994) 'Aesthetics and political economy in the eighteenth century: the place of gender in the social constitution of knowledge' in George Levine (ed.) *Aesthetics and Ideology*, New Brunswick, NJ: Rutgers University Press.

——(1994a) 'The social constitution of "class": toward a history of classificatory thinking' in Wai Chee Dimock and Michael T. Gilmore (eds) *Rethinking Class: Literary Studies and Social Formations*, New York: Columbia University Press.

——(1995) *Making a Social Body: British Cultural Formation, 1830–1864*, Chicago, IL: University of Chicago Press.

——(1998) *A History of the Modern Fact: Problems of Knowledge in the Sciences of Wealth and Society*, Chicago, IL and London: University of Chicago Press.

——(2008) *Genres of the Credit Economy: Mediating Value in Eighteenth- and Nineteenth-Century Britain*, Chicago, IL: University of Chicago Press.

Poulot, Dominique (1994) 'Identity as self-discovery: the eco-museum in France' in Daniel J. Sherman and Irit Rogoff (eds) *Museum Culture: Histories, Discourses, Spectacle*, Minneapolis, MN: University of Minnesota Press.

——(2005) *Une histoire des musées de France, XVII-XX siècle*, Paris: Éditions la Découverte.

Prakash, Gyan (2002) 'The colonial genealogy of society: community and political modernity in India' in Patrick Joyce (ed.) *The Social in Question: New Bearings in History and the Social Sciences*, London and New York: Routledge.

Preziosi, Donald (1996) 'In the temple of entelechy: the museum as evidentiary artifact' in Gwendolyn Wright (ed.) *The Formation of National Collections of Art and Archaeology*, Washington, DC: National Gallery of Art.

——(2003) *Brain of the Earth's Body: Art, Museums, and the Phantasms of Modernity*, Minneapolis, MN and London: University of Minnesota Press.

Price, Sally (2007) *Paris Primitive: Jacques Chirac's Museum on the Quai Branly*, Chicago, IL: University of Chicago Press.

Ravaisson, Félix (2008 [1838]) *Of Habit*, London and New York: Continuum.

Rancière, Jacques (1989) *The Nights of Labor: The Workers' Dream in Nineteenth-Century France*, Philadelphia, PA: Temple University Press.

——(1991) *The Ignorant Schoolmaster: Five Lessons in Intellectual Emancipation*, Stanford, CA: Stanford University Press.

——(1992) *The Names of History: On the Poetics of Knowledge*, Minneapolis, MN and London: University of Minnesota Press.

——(1999) *Disagreement: Politics and Philosophy*, Minneapolis, MN and London: University of Minnesota Press.

——(2002) 'The aesthetic revolution and its outcomes: emplotments of autonomy and heteronomy', *New Left Review*, 14, 133–52.

——(2004) *The Philosopher and his Poor*, Durham, NC and London: Duke University Press.

——(2004a) *The Politics of Aesthetics: The Distribution of the Sensible*, London and New York: Continuum.

——(2006) 'Thinking between disciplines: an aesthetics of knowledge', *Parrhesia*, 1, 1–12.

——(2009) *Aesthetics and its Discontents*, Cambridge: Polity.

——(2009a) 'A few remarks on the method of Jacques Rancière', *Parallax*, 15 (3), 114–23.

——(2011) 'The thinking of dissensus: politics and aesthetics' in Paul Bowman and Richard Stamp (eds) *Reading Rancière*, New York: Continuum.

Ray, William (2001) *The Logic of Culture: Authority and Identity in the Modern Era*, Oxford: Blackwell.

Rees-Leahy, Helen (2009) 'Assembling art, constructing heritage: buying and selling Titian, 1798 to 2008', *Journal of Cultural Economy*, 2 (1–2), 131–45.

Révai, Josef (1971) 'A review of Georg Lukács' *History and Class Consciousness*', *Theoretical Practice,* 1, 3ff.

Rexer, Lyle and Rachel Klein (1995) *American Museum of Natural History, 125 Years of Expedition and Discovery*, New York: H. N. Abrams and the American Museum of Natural History.

Ribot, Théodule Armand (1997 [1882]) *Diseases of the Will*, Washington, DC: University Publications of America.

Rivet, Paul and Georges-Henri Rivière (1933) 'La mission ethnographique et linguistique Dakar-Djibouti', *Minotaure*, 2, 3–6.

Rivet, Paul, Paul Lester and Georges Henri Rivière (1935) 'Le laboratoire d'anthropologie du Muséum', *Archives du Muséum d'Histoire naturelle,* 6th series, 12, 507–31.

Rivière, George Henri and Paul Rivet (1931) 'Rapport sur la réorganisation général du Musée', Archives of the Musée de l'Homme, 2 AM 1 G2b: *Dossier relative à la reorganisation générale du Musée d'Ethnographie du Trocadéro.*

Roberts, Nathan (2004) 'Character in the mind: citizenship, education and psychology in Britain, 1880–1914', *History of Education*, 33 (2), 177–97.

Rose, Jonathan (2002) *The Intellectual Life of the British Working Class*, New Haven, CT: Yale University Press.

Rose, Nikolas (1985) *The Psychological Complex: Psychology, Politics and Society in England, 1869–1939*, London: Routledge and Kegan Paul.

——(1996) 'The death of the social? Refiguring the territory of government', *Economy and Society*, 25 (3), 327–56.

——(1996a) 'Authority and the genealogy of subjectivity' in Paul Heelas, Scott Lash and Paul Morris (eds) *Detraditionalisation: Critical Reflections on Authority and Identity*, Oxford: Blackwell.

——(1998) *Inventing Ourselves: Psychology, Power, and Personhood*, Cambridge: Cambridge University Press.

——(1999) *Powers of Freedom: Reframing Political Thought*, Cambridge: Cambridge University Press.

Ross, Kristin (1991 'Rancière, and the practice of equality', *Social Text*, 29, 57–71.

Rowse, Tim (1992) *Remote Possibilities: The Aboriginal Domain and the Administrative Imagination*, Casuarina: North Australian Research Unit, Australian National University.

——(1996) 'Rationing the inexplicable' in S. R. Morton and Derek John Mulvaney (eds) *Exploring Central Australia: Society, Environment and the 1894 Expedition*, Chipping Nortin, NSW: Surrey Beatty.

——(1998) *White Flour, White Power: From Rations to Citizenship in Central Australia*, Cambridge: Cambridge University Press.

——(2009) 'The ontological politics of "closing the gaps"', *Journal of Cultural Economy*, 2 (1–2), 33–48.

Roy, Tania (2006) 'The aesthetic in colonial India', *Theory, Culture & Society*, 23 (2–3), 244–46.

Segalen, Martine (2005) *Vie d'un musée 1937–2005*, Paris: Éditions Stock.

Serres, Michel (1982) *The Parasite*, Baltimore, MD and London: Johns Hopkins University Press.

Shaftesbury, Third Earl of (Anthony Ashley Cooper) (1999) *Characteristics of Men, Manners, Opinions, Times*, Cambridge: Cambridge University Press.

Shapin, Steven and Simon Schaffer (1985) *Leviathan and the Air-Pump: Hobbes, Boyle and the Experimental Life*, Princeton, NJ: Princeton University Press.

Sherman, Daniel (2004) '"Peoples ethnographic": objects, museums, and the colonial inheritance of French ethnology', *French Historical Studies*, 27 (3), 669–703.

Sibeud, Emmanuelle (2007) 'The metamorphosis of ethnology in France, 1839–1930' in Henrika Kuklick (ed.) *A New History of Anthropology*, Malden, MA and Oxford: Blackwell.

Singer, Brian C.J. (1999) 'Méditations pascaliennes: The *Skholè* and democracy', *European Journal of Social Theory*, 2(3), 282–97.

Skinner, Ghislaine M. (1986) 'Sir Henry Wellcome's Museum for the Science of History', *Medical History*, 30, 383–418.

Skinner, Quentin (2002) *Visions of Politics. Volume 2: Renaissance Virtues*, Cambridge: Cambridge University Press.

Smith, Adam (2002) *The Theory of Moral Sentiments*, Cambridge: Cambridge University Press.

Spencer, Baldwin (1914) 'The Aboriginals of Australia' in G. H. Knibbs (ed.) *Federal Handbook prepared in connection with the eighty-fourth meeting of The British Association for the Advancement of Science held in Australia*, August, Melbourne: Commonwealth of Australia.

——(1921) 'Blood and shade divisions of Australian tribes', *Proceedings of the Royal Society of Victoria*, New Series, part 1, 34, 2–6.

——(1922) *Guide to the Ethnological Collection Exhibited in the National Gallery of Victoria*, 3rd edition, Melbourne: Government Printers.

Spencer, Baldwin and Frank J. Gillen (1899) *The Native Tribes of Central Australia*, London: Macmillan & Co.

——(1912) *Across Australia*, vol 1, London: Macmillan and Co.

Spencer, Frank (1992) 'Some notes on the attempt to apply photography to anthropometry during the second half of the nineteenth century' in Elizabeth Edwards (ed.) *Anthropology and Photography 1860–1920*, New Haven, CT and London: Yale University Press in association with the Royal Anthropological Institute.

Spencer, Herbert (1996 [1855]) *The Principles of Psychology*, London: Routledge/Thoemmes Press.

Spieker, Sven (2008) *The Big Archive: Art from Bureaucracy*, Cambridge, MA: MIT Press.

Steinmetz, George (2011) 'Bourdieu, historicity, and historical sociology', *Cultural Sociology*, 5 (1), 45–66.

Stocking, George W., Jr. (1983) 'The ethnographer's magic: fieldwork in British anthropology from Tylor to Malinowski' in George W. Stocking, Jr. (ed.) *Observers Observed: Essays on Ethnographic Fieldwork*, Madison, WI: University of Wisconsin Press.

——(1987) *Victorian Anthropology*, New York: Free Press.

——(1991) 'Maclay, Kubary, Malinowski: archetypes from the dreamtime of anthropology' in George W. Stocking, Jr. (ed.) *Colonial Situations: Essays on the Contextualisation of Ethnographic Knowledge*, Madison, WI: University of Wisconsin Press.

Summit, Jennifer (2008) *Memory's Library: Medieval Books in Early Modern England*, Chicago, IL and London: University of Chicago Press.

Thoburn, Nicholas (2002) 'Difference in Marx: the lumpenproletariat and the proletarian unnamable', *Economy and Society*, 31 (3), 434–60.

Thomas, David Wayne (2004) *Cultivating Victorians: Liberal Culture and the Aesthetic*, Philadelphia, PA: University of Pennsylvania Press.

Thompson, Grahame F. (2003) *Between Hierarchies and Markets: The Logic and Limits of Network Forms of Organisation*, Oxford: Oxford University Press.

Toews, David (2009) 'The new Tarde: sociology after the end of the social,' *Theory, Culture & Society*, 20 (5), 81–98.

Toscano, Albert (2011) 'Anti-sociology and its limits' in Paul Bowman and Richard Stamp (eds) *Reading Rancière*, New York: Continuum.

Tulley, James (1989) 'Governing conduct: Locke on the reform of thought and behaviour' in Edmund Leites (ed.) *Conscience and Casuistry in Early Modern Europe*, Cambridge: Cambridge University Press.

Tylor, Edward Burnett (1867) 'On traces of the early mental condition of man', *Notices of the Proceedings at the Meetings of the Royal Institution of Great Britain*, 5, 83–93.

——(1867a) 'Phenomena of higher civilisation: traceable to a rudimentary origin among savage tribes', *Anthropological Review*, 5 (18/19), 303–14.

——(1869) 'On the survival of savage thought in modern civilisation', *Notes on the Proceedings at the Meetings of the Royal Institute*, 5, 522–35.

——(1871) *Primitive Culture*, 2 vols, London: John Murray.

Umbach, Maiken (2009) *German Cities and Bourgeois Modernism, 1890–1924*, Oxford: Oxford University Press.

Urry, John (2000) *Sociology Beyond Societies*, London: Routledge.

Valverde, Mariana (1996) '"Despotism" and ethical liberal governance', *Economy and Society*, 25 (3), 357–72.

——(1998) *Diseases of the Will: Alcohol and the Dilemmas of Freedom*, Cambridge: Cambridge University Press.

Wasson, Haidee (2005) *Museum Movies: The Museum of Modern Art and the Birth of Art Cinema*, Berkeley, Los Angeles and London: University of California Press.

——(2008) 'Studying movies at the museum: the Museum of Modern Art and cinema's changing object' in Lee Grieveson and Haidee Wasson (eds) *Inventing Film Studies*, Durham, NC and London: Duke University Press.

Weber, Florence (2000) 'Le folklore, l'histoire, et l'état en France (1937–45)', *Revue de synthèse*, 4th series, 3–4, 453–67.

Weber, Max (2001) *The Protestant Ethic and the Spirit of Capitalism*, London: Routledge.

White, Hayden (1992) 'Foreword: Rancière's revisionism' in Jacques Rancière, *The Names of History: On the Poetics of Knowledge*, Minneapolis, MN and London: University of Minnesota Press.

White, Melanie (2005) 'The liberal character of ethnological governance', *Economy and Society*, 34 (3), 474–94.

Whitehead, Christopher (2009) *Museums and the Construction of Disciplines: Art and Archaeology in Nineteenth-Century Britain*, London: Duckworth.

Wilder, Gary (2003) 'Colonial ethnology and political rationality in French West Africa', *History and Anthropology*, 14 (3), 219–52.

——(2005) *The French Imperial Nation-State: Negritude and Colonial Humanism between the Two World Wars*, Chicago, IL and London: University of Chicago Press.

Willis, Paul (1977) *Learning to Labour: How Working-Class Kids Get Working-Class Jobs*, New York: Columbia University.

Wolfe, Cary (2010) *What is Posthumanism?* Minneapolis, MN and London: University of Minnesota Press.

Wolfe, Patrick (1991) 'On being woken up: the Dreamtime in anthropology and in Australian settler culture', *Comparative Studies in Society and History*, 33 (2), 197–224.

——(1999) *Settler Colonialism and the Transformation of Anthropology: The Politics and Poetics of an Ethnographic Event*, London: Cassell.

Wolff, Christian (1750) *The Real Happiness of a People under a Philosophical King*, London: M. Cooper.

Woodmansee, Martha (1994) *The Author, Art, and the Market: Rereading the History of Aesthetics*, New York: Columbia University Press

Yúdice, George (2003) *The Expediency of Culture: Uses of Culture in the Global Era*, Durham, NC and London: Duke University Press.

Index